Theory of Science

An Introduction to the History, Logic, and Philosophy of Science

Theory of Science

An Introduction to the History, Logic, and Philosophy of Science

George Gale
University of Missouri, Kansas City

McGraw-Hill Book Company
New York St. Louis San Francisco Auckland Bogotá Düsseldorf
Johannesburg London Madrid Mexico Montreal New Delhi Panama
Paris São Paulo Singapore Sydney Tokyo Toronto

THEORY OF SCIENCE
An Introduction to the History, Logic, and Philosophy of Science

1 2 3 4 5 6 7 8 9 0 D O D O 7 8 3 2 1 0 9 8

This book was set in Times Roman by Black Dot, Inc. The editors were Richard R. Wright and James R. Belser; the cover was designed by John Hite; the production supervisor was Dominick Petrellese. The drawings were done by ECL Art Associates, Inc.
R. R. Donnelley & Sons Company was printer and binder.

Library of Congress Cataloging in Publication Data

Gale, George, date
Theory of science.

Bibliography: p.
Includes index.
1. Science—Philosophy. 2. Science—History.
I. Title.
Q175.G23 501 78-16670
ISBN 0-07-022680-6

An awful lot of people got me interested in learning things. Then they, my teachers, taught me things to learn. This book is their fault, and I gleefully and thankfully dedicate it to each of them. Perhaps they will be more careful next time.

Contents

Introduction

Why would anyone want to investigate science in the first place? Science is well known to be the major activity of research and investigation in modern Western society. But with increasing speed, science itself is becoming the object of research and investigation. This turnabout is a relatively recent phenomenon; it is only during the last several decades that courses in the history, logic, and philosophy of science have more and more come to be offered in colleges and universities. There are many reasons behind this phenomenon, not the least of which is the growing significance of science and technology to our society. However, in addition to this, it seems to me that there are other important reasons, ones which are related to the benefits that students get from studying science itself.

The first reason concerns science majors, those students studying physics, or psychology, or medicine, or another science, as their main subject. For these students, an introduction to the history, logic, and philosophy of science provides a peculiar kind of counterbalance to the heavy doses of purely scientific data they receive in their major classes. Many scientists argue that such a counterbalance is vitally necessary to the continuation of a sane and healthy scientific tradition. Their reasoning goes like this: Scientific research is currently accumulating new data at an ever-accelerating pace. But all this new data, plus the fundamentally sound old data, must be presented within the confining time span of the normal four-year undergraduate career. In order to accomplish this, the data have to be presented in an increasingly more efficient manner. Most often what happens in these circumstances is that teaching of an essential element of scientific method is sacrificed in the move toward efficiency. The essential element that students do not become acquainted with is the notion that scientific results are ultimately reached through dialogue, through the ongoing give and take that occurs during public presentation and criticism of theories and experiments. According to this fundamental concept of scientific method, *every* piece of scientific data, *every* element of scientific theory is, in principle, subject to criticism and rebuttal; thus, every piece and element, in principle, could be replaced by an alternate view. This trial-by-fire process is the most central feature of modern science since scientists come to accept a new view or piece of data only if it survives the ordeal. However, because of the need to streamline their courses and make them more efficient, scientists/teachers often have not been able to make time available for consideration of the alternatives to, or competition of, currently accepted and prevailing views. Prevailing views thus tend to become treated as fact, as undeniable truth; in short, current scientific positions come to be presented as dogma.

I am sure that everyone is familiar with this aspect of science education. But it has its dangers. For example, I remember quite well my own shock and righteous outrage when a good friend in college began to ask me some very tough questions about electrons and their existence. As the questions got more and more difficult to deal with, I finally could respond only in the fashion of a "true believer": I treated my friend as if he were a heretic, a turncoat to the cause of science and truth. After this frustrating debate, I went to the library and looked up some books my friend had recommended. To my growing chagrin I came to realize that there really *were* some problems with electrons, and in fact, it soon became clear that some very reasonable men—even a physicist or two—did not believe in the existence of the electron. These individuals believed that an inventory list of the existing things of the universe would include, for example, tables and chairs, fleas and tissue cells, but it would not include unicorns and electrons. Given this realization, I first blamed my science teachers for pulling the wool over my eyes, for concealing the fact that there was an alternative view regarding the existence of electrons and other submicroscopic particles. But I later came to understand that it was the pressing demand of the four-year undergraduate science curriculum which was responsible for this deficiency in my education, and that my science teachers had not been in some vile conspiracy to hide the facts from me. By now you might very reasonably be wanting to ask, "So what? What is the point of all this?" My point is a very straightforward one. Modern science courses tend to teach only the prevailing view. Students cannot help coming to believe the prevailing view as dogma. Thus, they are much less open to consider views which run counter to the prevailing view. But closed minds are deadly to the progress of science. Even a quick glance at the history of science shows that almost all the great advances in theory and experiment were initially proposed as alternatives counter to the prevailing view and that each met with fierce intellectual resistance. For example, Lavoisier's proposal for oxygen theory, a radical alternative to the prevailing view, initiated a period of vigorous conceptual battle in chemistry. So did Copernicus's theory in astronomy, Pasteur's and Darwin's theories in biology, and Freud's hypothesis in psychology. The list goes on and on. In light of these, and other typical examples, my point is that perhaps we can lessen this intellectual resistance which is one result of turning out dogmatic scientists from our schools. Thus, in the hope that modern science can be spared the trials and tribulations of difficult conceptual transitional periods such as those of the past, and in the hope that competition among alternative views can be encouraged, and finally, in the hope that an effective counterbalance can be provided against the growing tendency to teach scientific views as dogma, more and more courses in the history, logic, and philosophy of science are being offered. The belief is that such courses can effectively influence science students.

Of course, the effect of such courses is also extremely relevant to nonscience students in that it provides them with a more comprehensive and accurate view of science as an historical institution and a tradition in Western culture. For the humanities major, or the fine arts major, or any other nonscience major, the typically brief undergraduate exposure to three or six or nine hours of science as a breadth requirement can only have resulted in a dogmatic view of science. Moreover, such breadth courses usually do absolutely nothing to enlighten the student about the role and status of science in contemporary civilization. But even more importantly, the student is left completely in the dark about what the nature of modern science *really* is (and how it got that way). Such lack of understanding certainly does not lead to closer relations between science and the humanities and arts, nor does it prepare the student for the participatory duties of a citizen in a modern scientific and technological society.

These reasons alone should be sufficient to prod nonscience students into courses on the history, logic, and philosophy of science. But there is a further reason for the nonscience major to take such courses: Even if, for these students, science and scientists are to become their intellectual "opposition," they should realize the value of "knowing the enemy."

It seems to me that reasons of the sort mentioned here are what lie behind the recent surge of interest in turning the tables on science, in order to make science *itself* the research subject, rather than the other way around. But whether or not this is the correct interpretation of the contemporary university scene, achievement of some or all of the benefits I have discussed here is certainly the objective which motivates me to write this book.

But before we move into a discussion of the specific issues, let me hereby announce the coining of a new phrase. In my discussion above I have used the phrase "history, logic, and philosophy of science" as a catchall term to refer to the new types of courses which study science. This is a clumsy and overlong phrase, of which efficiency would seem to demand removal in favor of some replacement term. I offer *theory of science* as a replacement. I am sure that some will find this term abusive, since the history of science, the logic of science, and the philosophy of science are all well-staked-out territories of academic turf. However, there are good reasons to compound elements of all three under one rubric. In the first place, all of the presently distinct fields, in terms of their actual histories, have individually coalesced out of the same earlier and more general field. Second, there are logically necessary connections between the fields; for example, history cannot be done without presupposing some particular philosophical position, and so on. Thus, although each of the three elements may be distinguished from the others in terms of its relatively independent subject matter, it may not be logically separated from the others. This latter reason alone is sufficient to warrant that this new name, *theory of science,* can be used appropriately to refer to what goes on in this book.

Acknowledgments

A number of people helped me endure the writing of this book. Some even helped me write it. My wife Carol added her expertise about English. Bruce Bubacz reviewed Chapter 2. Chapter 8 benefited from the insights of Hank Frankel. Marjorie Grene and Carlo Giannoni both contributed support and guidance. John Urani and Judy Hemberger helped with physics and biology, respectively. Jim Belser, my editor at McGraw-Hill, kept me soothed at all the proper times. To all of these friends and colleagues, I must say "Vielen Dank!" I am sure that their collective sigh of relief at final completion of this book is just as great as mine. Nan Biersmith—you too can relax with my special thanks, and finally let your typewriter cool off!

I also must acknowledge the permission from the various presses to cite the material in the text. You will find, along with some of the citations, recommendations to procure certain of these works. I can't acknowledge them any more highly!

George Gale

Theory of Science

An Introduction to the History, Logic, and Philosophy of Science

Part One

Science and Philosophy

Philosophy is an ancient occupation. There are written records of human philosophizing which date back almost 3000 years. But science, on the other hand, is a relatively new element in the human intellectual tradition. Modern science did not get a good foothold established until the sixteenth century. Moreover, even apart from their widely separated birthdays, science and philosophy would seem at first glance to be so *different* from one another. Science, after all, is test tubes, atom smashers, and good hard facts. Philosophy, however, usually is understood to be the wise sayings of old graybeards. What possible connection would there be between two apparently so divergent activities? It is the task of the first three chapters to reveal the connection.

In Chapter 1, philosophy and three of its main divisions—metaphysics, epistemology, and logic—are examined. Each of these examinations is carried out in terms of a scientific example. I try very hard to show that philosophical questions are quite different from scientific questions, even though at first they *sound* a lot like scientific questions. But it should soon become clear to you that these philosophical questions are just not the same as scientific questions. That is, they should soon reveal themselves as questions *about* science, and not as questions *within* science. Let me give you a quick example of what I mean here.

One significant element of modern science is the fact that it uses the experimental method of proof. Thus, a result such as "Atoms are composed of smaller particles such as neutrons and protons" counts as scientific—that is, *it is accepted as being scientific*—only if it has been tested and proved by experiment. If there have been no experiments, then it is not scientific. Having been subjected to experiment is a large part of what we mean when we call something "scientific." If we think of science as a special sort of game, then using experimental proof is one of the fundamental rules of the game. But we can ask "How did this rule get selected in the first place?" If science is a game, then the game cannot be played until the rules have been chosen. One does not first make up a game, and *then* choose the rules. The game cannot be made up until decisions about the rules have been made. Thus the question "Which method of proof shall we use in science?" is not a scientific question, even though it sounds like one. In fact, it is a philosophical question. Chapter 1 raises three of these scientific-sounding-but-not-scientific questions: (1) What is the world like at its most fundamental level; does it come in chunks like "atoms," or does it come instead in fuzzy volumes like "fields"? (2) When should scientific experiments be conducted—*before* the scientist reaches those tentative conclusions called "hypotheses," or only *after* these hypotheses have been proposed? (3) Can the most fundamental of all scientific proposals be subjected to experimental proofs if, in fact, the experiments themselves depend upon these principles?

Each of these questions is shown to belong to a specific part of philosophy. The first, for example, is identified as a "metaphysical" question, the second as an "epistemological" question, and the third as a "logical" question. (These words, and others that may be unfamiliar, are defined in the glossary at the end of the book.)

Following these three discussions in the first chapter, Chapter 2 gets down to a small piece of technical business. There is a language which has been developed to represent some aspects of scientific thinking. This language is called *symbolic logic.* I describe some basic elements of logic and point out how it can be used to make some clarifications about science. The language is relatively straightforward and should not present any big hurdle for you to comprehend.

Chapter 3 finally gets us down to some philosophical business. It is the job of the philosopher, the historian, and the scientist together to define what the notion "science" means and refers to. That is what goes on in Chapter 3. I examine some of the activities of scientists I know, consider some historical facets which are evident in the scientific tradition, and put these through a philosophical analysis. What emerges from this process are some proposals about science; for example, the notion that science exists not only to make our everyday life easier and more comfortable, but also simply to satisfy that distinctively human drive—curiosity! It also comes to light that there exist scientific traditions which are similar in many ways to political or religious traditions, although in other ways these types of traditions differ significantly.

In the final section of Chapter 3 I examine one very special scientific tradition. By the end of the discussion you will know why we sometimes say, "He sure is in a bad humor today" or "Mediterranean people are fiery and hot-blooded."

As you can see, these three chapters cover an enormous amount of territory. But the territory is fascinating to me, and I hope that I have described it well enough for it to grab your interest too.

Chapter 1

Introduction to Philosophy

PHILOSOPHY

The theory of science makes heavy use of concepts developed within the general subject matter called "philosophy." Although philosophy and philosophers have been around in Western civilization for over 2000 years, there still is not total agreement about what does and what does not precisely constitute philosophy. The term "philosophy" is found to be used for such mundane things as, say, Joe Namath's "philosophy of the forward pass" or Nelson Rockefeller's "philosophy of investment." But at the same time, the term is also used to include more exotic ideas such as Jean Paul Sartre's "philosophy of nausea" or Albert Einstein's "philosophy of relativity." However, even across this wide diversity of uses we can see a common thread: *Philosophy* seems to refer to the underlying set of reasons and concepts, that is, to the underlying theoretical foundations, of some activity. On this interpretation, Mary's or Bill's "philosophy of living" would be her or his ultimate set of concepts, the theory about how life was to be lived, about its purpose, its origin, and so on. In what follows I intend generally to preserve this underlying meaning of the term, although I will modify its use somewhat, according to the following definition: *Philosophy* is that *ultimate* system of concepts which people refer to in explanations of the events and occurrences of their lives.

Let me give an example of what I mean here, and at the same time introduce you to a new distinction. Let us suppose that we have taken a time machine trip back into ancient Greece. Further, suppose we are observing a parent and child out walking together, when a sudden lightning stroke and its accompanying thunderclap occur. The child asks, as children are wont to do, "What's that?" The parent answers, "Lightning and thunder." This response illustrates the first level of explanation. That is, when one is asked "What's that?" one simply answers by giving the name of the item in question. But children, like philosophers (or vice versa), are often unsatisfied with simply knowing the names of the occurrences in question. Children almost invariably push their questions ever more deeply into the issue at hand, it is this depth which eventually reveals the underlying philosophy of the person being questioned. Thus, we observe the child continuing to interrogate his parent. After having the names "thunder" and "lightning" explained to him, the child now asks, "But what *are* thunder and lightning?" The parent, with a bit of exasperation, answers, "Well, they are movements of the upper air." The child, again unsatisfied with the explanation even though this response has gone to a second, deeper level, asks further, "But what sets off the movements?" The even more exasperated parent now answers, "The gods are having a quarrel and hurling magic bolts at one another. That's what causes the air to move." But the child, again unsatisfied, asks "Why?" once more. At this point the out-of-patience parent answers only: "I don't know. I guess it's just because, that's why."

The point at which the ancient Greek parent in our situation (or any other person being questioned in similar circumstances) cannot come up with further underlying concepts of explanation is the point which reveals the starting point of the philosophy of the culture. Once an "I don't know . . . just because" is reached, reflection and further thought will begin to work out a philosophy. Thus, the "I don't know" occurs at the level of philosophy's first beginning. We see here in this particular example that there are three levels of explanation: first, the level on which merely the names of the events are given; second, the level which gives an immediate explanation of the events named, namely, "movements of the air"; and finally, the ultimate philosophical level. These three are layered together to make the total explanation. The explanation terminates, after reflection and thought have produced them, in philosophical concepts, that is, concepts which are fundamental in the sense that they can be produced only at the level of, "I don't know. I guess it's just because, that's why."

Different cultures and different eras produce different explanations, and thus reveal different philosophies. For example, let us tune in on another parent and child situation, again with attendant thunder and lightning. This time, however, we are observing a modern situation. After being told the names of the occurrences, the child of course asks, "But what *are* thunder and lightning?" The parent tells her daughter, "They are various kinds of movements in the air." The child, unsatisfied as usual, asks "But what sets off the movements?" The mother, who is a physicist, explains further, "The clouds and the earth have different charges, and when the difference in charge gets too great, they come together. The coming together is the lightning, and the air is so agitated by the lightning that it produces the thunder. That's what sets off the movements." The child is still not satisfied. She again asks for an explanation: "But why are there different charges?" At this point the mother is exasperated and can only answer, "I don't know. I guess it's just because, that's why."

Here again we find a series of requests for an ever deeper explanation. And again the explanation terminates in an ultimate set of concepts, which can only be terminated by, "I don't know. I guess it's just because, that's why." Philosophy is simply this ultimate level of concepts, the concepts which, in both examples, follow reflection at the level of the "I don't know. I guess it's just because, that's why" response. The role of the philosophical element in the two sets of explanations is logically identical, even though hundreds of human generations separate the individuals concerned.

But even given the identical logical role of the philosophical concepts, there are differences in the two explanations. The Greek explanation terminated in philosophical concepts which did not refer to everyday, observable, experiential objects and events of the natural world. Gods, to put it bluntly, are supernatural; they are not elements of the natural world in the same way that thunderclaps and air movements are.

On the other hand, electrical charges on clouds and earth are natural, observable events in a way in which gods are not. We distinguish between the two different types of philosophical concepts represented here in the two examples by calling the one type "naturalistic" and the other type "nonnaturalistic."

The distinction between naturalistic and nonnaturalistic is not the same distinction as that between "old-fashioned" and "modern." For example, we find naturalistic explanations in the philosophy of Aristotle in ancient Greece; and it is easy to imagine our modern parent going one step further in her explanation. When asked about the reason for the difference in electrical charges, she might respond, "Because that's the way God wanted it to be." Clearly this last step is a nonnaturalistic explanation.

The distinction between the two types of philosophical systems is not terribly difficult to see. Other examples of the difference come fairly easily to mind. Consider someone ill with a fever. One culture might ultimately explain the fever in terms of some identifiable bacteria which were found in the water supply. Another culture might respond to a request for explanation of the fever in terms of a spirit or devil which had come to inhabit the sick individual. A third culture might blend both sorts of ideas. But ultimately, no matter how many levels are gone through, no matter how elaborate the explanation is, it must terminate in some ultimate response. There always must be the "Just because, that's why" answer to the questioner. And it is at that level that the culture reveals its philosophy.

However, after saying all this, I do not want to give you the impression that it is always possible to make an absolutely clear-cut distinction between "naturalistic" and "nonnaturalistic" when it comes to some actual cases. For instance, some societies have used explanations which appear to be sort of halfway between the two poles. Thus, we might find "demigods" functioning in explanations; these creatures are conceived to be halfway between heaven and earth. At other times, various cultures have used *both* types of explanation at one and the same time. Here we would notice that the nonnaturalistic concept and the naturalistic concept were operating in parallel. But even given these complicated situations, the distinction between "naturalistic" and "nonnaturalistic" is a valuable one which we can (and will) use to clarify some of these otherwise messy situations.

The relevance of all this to the theory of science is fairly straightforward. Science is (and has been) considered by most of its investigators to be a community of individuals engaged in similar and related activities of inquiry. It is clear that there are some difficulties with this conception, but it seems to me to be so valuable that I must use it in spite of the problems with defining precisely how science is a community of inquirers. You should be able now to see the implication of all this discussion about philosophy: Insofar as science is itself a community of inquirers, then it must necessarily have a philosophy, just like any other group of people asking questions. I believe that this is true, and I hope to be able to convince you of the same. I am not claiming here, in my notion of a "community," that there are not disagreements, and we will spend quite a bit of time analyzing some of them. But I want to emphasize to you that these disagreements are *philosophical* conflicts—conflicts, among other things, in regard to ultimate kinds of explanatory ideas, ideas which lie just on the edge of the abyss of the "Just because, that's why" response. Let me now begin to introduce you to some more details about the terminology and notions of philosophy.

METAPHYSICS

The collected works of Aristotle are divided into discussions about various subjects. There are books about meteorology, logic, biology, physics, ethics, and so on. The book which immediately follows the book about physics is called "Metaphysics," meaning "after physics." This book is a study of the various kinds of things which may be said to ultimately exist. In his research, Aristotle analyzes all those things which both his ancestors and his contemporaries had declared to be ultimately real, including gods, matter, spirit, fire, and atoms. Aristotle accepts certain of these findings, rejects others, and discovers concepts of his own. Two of his concepts, "matter" (the stuff of which something is made) and "form" (the pattern or arrangements of the stuff), persist today in many philosophical systems, including that of the Roman Catholic Church. But the importance of Aristotle's work for our present topic is that the name "metaphysics" came to be used for any and all attempts to answer the question, "What are the fundamentally real objects of the universe?"

An enormous range of possible answers can be given to this question. Many different answers in fact have been given in human history. A clear division between two types of answers can be made in the terms introduced in the preceding section. That is, the question, "What objects are ultimately real?" may be answered naturalistically or nonnaturalistically. Thus, "divine spirits" might be held to be ultimately real in a philosophy which is nonnaturalistic. On the other hand, "material objects" might be held to be ultimately real within a naturalistic philosophy.

But naturalistic philosophies can also differ from one another. Some of the important disagreements between scientists have occurred because of different beliefs regarding alternative naturalistic answers to the metaphysical question. One significant example of such a conflict occurred early in the era of modern science, during the period A.D. 1700–1715. The scene of the controversy ranged across Europe and England. The major participants were two philosopher/scientists, Gottfried Leibniz on the Continent and Isaac Newton in England. Leibniz and Newton had both been independently involved in the creation of the mathematical technique known as the *calculus.* Calculus involves analysis of things like velocity and acceleration over smaller and smaller intervals of space and time, until the mathematical difference between two succeeding intervals tends toward zero, that is, toward an infinitesimally small difference. The value of the technique results from the fact that precision grows as the interval of difference shrinks. Each of the two scientists had done his own independent work on the new technique, and had found various experiments which allowed interpretation in terms of the calculus. But in general, the two men tended toward significantly different interpretations of the meaning and significance of the technique of the calculus.

Newton believed that the success of the calculus demonstrated that the universe was ultimately composed of infinitesimally small point-masses, i.e., objects or particles ("atoms") which were vanishingly small, and which had vanishingly small mass.[1] Moreover, he also believed that the very large objects of the universe could be approached in terms of their point-masses. For example, Newton demonstrated that the gravitational centers of the planets and sun were point-masses which corresponded to the infinitesimals of the calculus. By this way of conceiving things, he was able to demonstrate the accuracy of his theory of universal gravitation. His theory, for the first time in history, allowed highly accurate predictions about the orbital trajectories of the planets and satellites. The quantitative success of Newton's theory far surpassed that of any of his predecessors or contemporaries; in large amount, it was this which vaulted the particulate or "atomistic" metaphysics into the forefront of scientific philosophy. Thus, following in the footsteps of his scientific/philosophical predecessors, Thomas Hobbes and Robert Boyle, Newton publicly adopted the metaphysics of atomism, and proclaimed a belief that the ultimate objects of the universe, the objects which underlay and constituted our everyday objects such as tables and chairs, were infinitesimally small, round, hard entities which were called, using the terminology of the ancient Greek philosophers, *atoms.* It was these objects which Newton conceived to correspond to the mathematical objects he knew as infinitesimals. Thus, in his view, his mathematical depiction of reality corresponded to reality itself, since both reality and its mathematical depiction contained infinitesimally small objects. A Newtonian explanation, then, for something such as light rays and their behavior, would involve the resolution of the observable properties of the light rays (for example, their intensity and direction) into properties of their underlying constituent particles. Using this conception, Newton hypothesized that reflection of light off a mirror was in reality the bouncing of a large number of tiny light particles from the surface. In the final analysis, the atomistic metaphysics requires that explanations of all of our everyday objects are to be resolved into their constituent parts, the particles or atoms which underlie them. Thus, in the Newtonian philosophy, the answer to the metaphysical question "What objects are ultimately real?" is the familiar one: "Atoms, or fundamental particles, are the ultimate objects of the universe."

Although Newton's metaphysics has endured in large part into our own day and age, it is not without competition today, nor was it without

[1]Sir Isaac Newton, *Opticks* (New York: Dover Publications, Inc., 1952), query 31, p. 400.

competition even in Newton's own time. Leibniz, and others of his Continental colleagues, felt that Newton had misinterpreted the operations of the calculus. Leibniz believed that the process of taking smaller and smaller intervals implied that there was no least-small interval; thus, he believed that there did not exist any lower limit to how small a space or a time might be. To him, it seemed that Newton's belief in fundamental particles involved the idea that there was some least-small piece of space and matter, namely, the atom itself. Leibniz argued that, if atoms existed, then they themselves, just like the infinitesimal mathematical objects of the calculus, must be capable of being divided, at least in our minds. This implies that atoms are not really ultimate, but rather, that the parts of atoms, the chunks into which we have divided them, are in fact really ultimate. Moreover, the chunks themselves may be subdivided, and so on ad infinitum. For this reason, Leibniz denied that there were any least-small objects in the world; in other words, he denied that there were any ultimate particles, or atoms.

But obviously Leibniz and his colleagues must have had *some* belief about the ultimate stuff of the universe, besides merely denying that Newton's conception was correct. In fact, Leibniz held a belief which has persisted with increasing value into our own day.[2] He believed that ordinary, observable objects were not divisible into ultimate parts but rather, that ordinary objects were the manifestation of underlying collections of forces. Moreover, these forces, although they had centers, were extended in volumes of influence throughout the entire universe. In later works, these volumes of force came to be called "fields" by men such as Michael Faraday and Albert Einstein. Leibniz thus denied Newton's atomistic conception of material objects, in favor of a notion of forces extending throughout space in fields. The fields, of course, unlike the atoms, could be infinitely subdivided.

We see in this conflict between Leibniz and Newton a metaphysical disagreement about what are the ultimately real objects in the universe. Newton believes that we can continuously subdivide material objects until we reach some fundamental level of particles. Leibniz, on the other hand, believes that we can continue forever to subdivide material objects and that we will never reach an end of the division process because matter is not divided into particles, but rather, is the result of the interplay of fields of force.

A fundamental metaphysical disagreement such as this is bound to have ramifications for other aspects of the sciences concerned. This became immediately obvious in the case of Leibniz and Newton. Leibniz's metaphysics proved to be more conducive to theories involving

[2]See George Gale, "Leibniz and Some Aspects of Field Dynamics," in *Studia Leibnitiana*, vol. VI, no. 1 (1974), p. 28.

dynamics and energy. Thus, his own scientific work culminated in the proposal that *vis viva* (the force notion which later developed into our modern idea of kinetic energy $= 1/2\ mv^2$) was always conserved in the universe. Moreover, field phenomena such as electricity and magnetism came to be more at home within metaphysical systems similar to that of Leibniz. On the other hand, Newton's metaphysics also had its beneficiaries. For example, kinetic theories such as the kinetic theory of gases—in which the gas is taken to be composed of large numbers of infinitesimally small particles—found a comfortable conceptual welcome within the atomistic ideas favored by Newton and his colleagues.

In contemporary science we still find the strains and tensions between these two fundamentally opposed metaphysical views.[3] Thus, in a peculiar way, the question whether light is a wave or a particle can be referred back to the conflict between Leibniz and Newton over which metaphysics was more appropriate to physics. It can be seen from this example that metaphysical questions are significant ones for science and for the scientists who are attempting to push their conceptions beyond what is familiar and into new zones of the natural world. In later chapters I will introduce you to several other metaphysical puzzles.

EPISTEMOLOGY

When people disagree philosophically, they do not always disagree about metaphysics. Instead they might disagree about *epistemology.* This term refers to what has been called "theory of knowledge." What it comes down to is this: If there can be thought to be different kinds of ultimate objects in the universe, then there must be different ways of coming to know these objects. A simple example might make my point more clearly. Consider two persons, one a naturalist philosopher and the other a nonnaturalist philosopher. The metaphysical system of the first person concludes with the belief that material objects, such as atoms or force fields or energy packets, are ultimately real. The metaphysical system of the second person concludes with the belief that a personal God is ultimately real. These divergent metaphysical beliefs necessarily imply differences in belief about how human beings can best come to know and understand the fundamental metaphysical object(s). The naturalist, for example, since the ultimate objects are natural ones, believes that the natural processes of seeing, hearing, touching—in short, the natural processes of sensation and perception—are the proper routes toward knowledge of the ultimate objects.

[3]See George Gale, "Forces and Particles: Concepts Again in Conflict" in *Journal of College Science Teaching,* vol. III, no. 1 (October 1973), p. 29.

The nonnaturalist, on the other hand, believes in an object which transcends everyday sensation and perception. Thus, knowledge of God must come from some process other than seeing, hearing, touching, etc. In some cases in history, this transcendent process has been one of meditation; in others it has involved personal dialogue via miracles, and so on. But the particular methodology is not what is important here; what is important is simply to notice that just as people differ in their metaphysics, they can and usually do differ in regard to what they consider the most appropriate method for knowing and comprehending the objects implied by their metaphysical beliefs. Thus, epistemological differences go hand in hand with metaphysical differences.

One major epistemological divergence has been present in science since the time of the classical Greek scientists Pythagoras and Aristotle. Among other things, Pythagoras strongly believed that pure mathematical thought would produce knowledge about the natural world. Aristotle, on the other hand, believed that scientific knowledge in biology, for example, would come through careful observation of actual specimens. Although neither of these men held views terribly similar to the views of modern scientists, still the emphases I note here exemplify a real difference in belief. Thus, the divergence in views exhibited in these two men's epistemologies can still be found today among major scientists. Let me just give one example.

The typical modern view about scientific methods, a view we are all familiar with, is that the scientist first sets up an experiment; second, observes what occurs in the experiment; third, reaches a preliminary hypothesis to describe the occurrence; fourth, runs further experiments to test the hypothesis; and finally, corrects or modifies the hypothesis in light of the results of the extended experimental test. This interpretation of scientific method has been a popular one, and indeed, it has been the orthodox interpretation in American methodology courses for several decades. Versions roughly similar to the one I outline here can be found even in textbooks used in elementary schools. Although this seems a purely modern view, its history is longer than you might expect. As I noted, a roughly parallel proposal that scientific knowledge was to be gained only through observation can be found in Aristotle's discussions of biological science. But the significant modern statement of this view was made first in the writing of Francis Bacon (c. 1620).[4] Bacon announced to the general public details of the "new experimental philosophy" which had been coming into use in the work of European scientists, most particularly Galileo. Although Galileo himself had described the require-

[4]Francis Bacon, "Novum Organum," in J. M. Robertson (ed.), *Philosophical Works*, after the text and translation of R. L. Ellis and J. Spedding (London: 1904).

ments of the new experimental epistemology, his works at this time were not immediately accessible to the general intellectual culture, especially in England. Bacon's role in this case was that of a promulgator. He argued that the value of the new kind of knowledge was great, especially in terms of its practical effects. Knowledge is power, Bacon claimed, power to make human life easier and more comfortable.

The experimental epistemology soon came into great favor among many philosophers both on the Continent and in England. The Baconian method came to be called *induction*; that is, experiment was thought to *induce* data fit for generalization. One mainstream of modern philosophy developed around the central idea that knowledge could come only from strict observation of empirical situations. This philosophical position was called *empiricism*, and in its strictest formulation (similar to that proposed by the philosopher David Hume), it considers scientific knowledge to be wholly and entirely limited to descriptions—observation statements, generalizations, and the like—which are developed from pure sensory experience. On this view, the scientist is limited to describing only what can be observed with the eyes, ears, nose, and so on. According to another philosopher, George Berkeley, the empiricist position in philosophy also has an attendent metaphysics: Something can be said to be real, to actually exist, only if it is an observable type of object or event. If an object or event cannot be conceived to be observable under any circumstances, then that object or event cannot be counted among the existent entities of the universe.

It is interesting that the empiricist position first developed an epistemology, a theory about what was to count as knowledge, and then went on to develop a metaphysical position which was consistent with its epistemology. Thus, the requirements for scientific knowledge acted as constraints upon the permissible metaphysical beliefs. Empiricism, or something similar to it, is still a healthy philosophical position among many of today's scientists. In particular, among this group of empiricists, the method of induction is afforded a central role as the only appropriate means which science can use to reach real knowledge. This stricture upon scientific method is especially evident among the Anglo-American social and behavioral scientists. Behaviorism, the epistemological view propounded by the American psychologists Watson and Skinner, although modified a great deal, is quite recognizable as a latter-day descendant of the modern empiricist epistemology initiated three and a half centuries ago by Bacon.

Empiricist epistemology in general, and induction in particular, is not, however, the only alternative theory of scientific knowledge. Another position, called *rationalism*, developed in competition with empiricism. Empiricism developed from the initial view that the human sensory system was the only legitimate starting point for scientific knowledge. In this view, it would not be stretching things too far to claim that, in a fundamental sense, the "experiment" (human experience) happened first, and scientific knowledge was distilled, induced as it were, from the experiment. But rationalism also developed in the emerging context of modern experimental science. Just as empiricism was one possible response, one possible construction of an epistemology congruent with modern experimental science, so rationalism was another. Rationalism, however, ended up with a quite different view of the role of experiment than did empiricism.

Rationalism acquired its name because it chose to emphasize the human mental power of reason (*ratio* in Latin), rather than the human mental power of empirical sensation, as did the empiricists. Rationalist scientists believed that it was possible, by pure unaided reason, first, to conceive and comprehend certain very general features of the universe, and then, from these conceptions, to deduce mathematically a description of what the actual empirical world was like, prior to any experiment. The role of experiment, in this interpretation of scientific method, would be as a decision procedure for testing between alternative deduced results. If one reasoned mathematically and came to the conclusion that x would be the actual situation in the world, then an experiment could be designed to check whether or not x really did occur.

Rationalism might sound strange to the modern educated mind. The strangeness has two reasons, as I understand the situation. In the first place, empiricism and rationalism are opposed beliefs about the appropriate method to reach true scientific knowledge. But empiricism of some sort is the usual epistemology which our modern educational institution places rather dogmatically before the student. Consequently, rationalism sounds like a strange and inappropriate epistemology for science according to anyone who has been dogmatically educated in empiricism. But this is only one of the reasons underlying the strangeness of the sound of rationalism. There is yet another, namely, that most modern educated minds are not now, and have never been, accurately introduced to the actual, real-life procedures of the contemporary *theoretical* (mathematical) scientist. The mathematical scientist, most typified by the theoretical physicist (although there are mathematical biologists, psychologists, chemists, and so on), follows none other than the rationalist method. But since the modern educated mind is not familiar with these sorts of contemporary scientists, it is also not familiar with rationalism. In an attempt to introduce you to this alternative epistemology, let me give you an example of rationalist science at work.

Leibniz, who was mentioned above in connection with the field versus particle metaphysical conflict, is a prime example of a scientist who believed, and proclaimed, the rationalist method of reaching scientific knowledge. One particular case of his has assumed the proportions of a classic instance of rationalism at work. Leibniz for various reasons (including among other things, his belief that God was an able and efficient architect) believed that Nature always acted in the optimum fashion. He expressed this belief in the idea that Nature always produced the greatest amount of effect for the least amount of expenditure in energy. This idea was why he had hypothesized the universal conservation of kinetic energy.

But Leibniz's belief in the optimality of Nature led to more than this very general rule about the physical universe. He went on to postulate very fine details about the behavior of physical systems. For example, he hypothesized that geometrical optics, the science which mathematically describes the paths of light rays as they reflect off surfaces and as they bend going through transparent materials, could be done in a rationalistic fashion according to the dictate that Nature always followed the optimal method in accomplishing her effects.[5] He visualized the following situation. Suppose that there is a light ray originating at point 1, just above the mirror, M. Further suppose that you desire to have the light ray end up at point 3 after it has bounced off the mirror. (See Fig. 1-1.) The question you then ask yourself is: If Nature follows the optimal pathway, what will be the location of the bounce point on the surface of the mirror? Empiricist epistemologists, of course, would already be frowning; "That's a funny question to ask," they would say. "Rather than ask it in that form, why don't you just turn on the light source and experiment until you find what is the location of the bounce point on the mirror?" But rationalist epistemologists would not be deterred by this criticism, since, like

[5]See George Gale, "Does Leibniz Have a Practical Philosophy of Science?; or, Does 'Least-work' Work?" in *Studia Leibnitiana*, suppl. 13, 1974, p. 151.

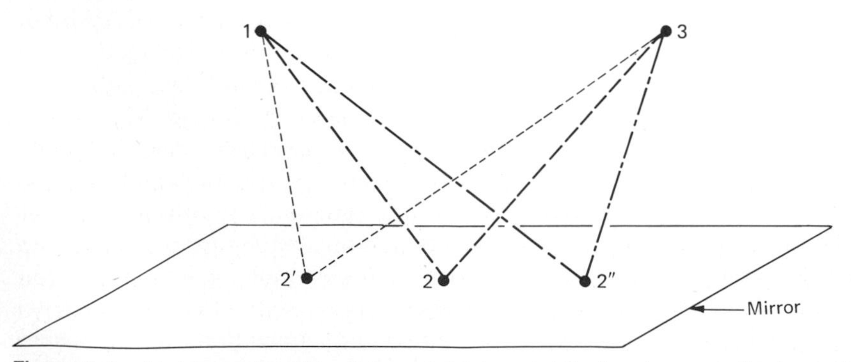

Figure 1-1. Light rays traveling from point 1 to point 3, via a bounce off mirror, M.

Leibniz, they have a firm belief that human reason can deduce prior to experiment what the location will be, rather than induce it from the data after the experiment. In his own case, Leibniz considered the fact that there were an infinite number of possible pathways from 1 to 3 via a mirror bounce. He pointed out, for example, there was path 1–2–3 and path 1–2′–3 and 1–2″–3, and so on, ad infinitum. But he deduced from considerations about optimality that the pathway would be 1–2–3, and no other in any circumstance. He then tested this prediction via experiment, and was shown to be completely correct. The process of deduction which he went through is absolutely fascinating. Let me tell you some of the details.

In the first place, he noted that path 1–2′–3 and 1–2″–3 are twins to one another. That is, their lengths and angles are the same, only the angles have been reversed in location. In general, just like these two, *all* paths between points 1 and 3 via the mirror will have identical twins. However, there is one exception to this general rule, namely path 1–2–3. This path does *not* have a twin—and it is the single solitary path which does not. Leibniz deduced from this fact one of the two reasons why light had to take the untwinned path. It is obvious, he claimed, that it takes less specific information, less "causality," for the light ray to take the untwinned path, since, for any of the twinned paths, Nature would have to build in additional determining factors in order to cause the light ray to take just one out of the two absolutely identical twins. Thus, he reasoned, for the light ray to take the untwinned path is the optimal outcome, since his theory assumes that Nature accomplishes her effects with the simplest sort of determining factors. But this reason was not the most compelling one he developed. The really significant one, to our minds as well as his, is this: The untwinned path is the *shortest* path between 1 and 3 via the mirror. Thus, the light ray gets there fastest, and with the least amount of effort and energy, by following the untwinned path. This path, to use a modern, and somewhat inappropriate but relevant phrase, is "the path of least resistance." (Leibniz had a third reason, related to the details of the calculus, which need not concern us here.)

It is important to note what is going on in this example. The rationalist epistemology has as its central claim the belief that human reason, independently of and prior to experiment, can reach scientific knowledge about the natural world. Experimental results are to function only as corroborative evidence of the truth and validity of the deductive statements reached by reason. In this example, Leibniz has deduced a prediction about the way the world behaves, and he apparently has reached this prediction prior to doing any experiment. Now it is clear that there are many fudge-factors built into his deduction. For instance, we would want very much to quiz him about his interpretation of the meaning of "optimal," since it is conceivable that this meaning changes when it enters new contexts. However, even given Leibniz's fudge-factors, it is clear that there is a marked difference between his view about what is the appropriate method for reaching scientific knowledge as compared with the view of his empiricist colleagues. Given these kinds of differences in epistemology, it is clear that scientists from the opposing camps will tend to set up their experiments somewhat differently and, moreover, will reason about their results in a different fashion.

At this point, you might be wondering whether or not there are still any practicing rationalists alive. Rigorous empiricism is still alive and well in its modern guise, "behaviorism." Rationalism, as I hinted at the outset of this present discussion, is indeed alive and kicking, especially in theoretical physics. One especially good example is to be found in the work of the British physicist P. A. M. Dirac.[6] Early in this century, before any experiments had been done to test the highly controversial quantum theory, Dirac announced to the world that he had completely accepted the basic equation of the theory. Dirac's acceptance demonstrated that he took the theory to be accurate scientific knowledge independent of experiment. He gave as his reason for acceptance of the theory the fact that he found the equations to be "beautiful" and "aesthetically elegant." The equations appealed directly to his rational mind, which had inspected the equations and deduced that they were appropriate descriptions of a Nature which was beautiful and (dare I say it?) optimal.

Another modern example can be found in the work of the American physicist Archibald Wheeler. Wheeler has declared his hunch that it may be possible to deduce the laws of Nature, that is, to reach ultimate scientific knowledge, purely from a mathematical consideration of the fundamental laws of formal logic.[7] According to his hypothesis, the empirical situation can be anticipated and described, prior to actual experiment, through the employment of pure reason. It would be difficult to find an example which more clearly exhibits the rationalist alternative to the empiricist view that scientific knowledge comes only after experiment and observation.

Perhaps the most delightful example of the contrast between these two opposed scientific philosophical views is to be found in an apocryphal story told about Albert Einstein's wife. The director of one of the world's largest and most complete experimental laboratories extended to Mrs. Einstein an invitation to tour their site. She came one day, and was escorted through acre after acre of huge, complicated, and obviously

[6]P. A. M. Dirac, "The Evolution of the Physicist's Picture of Nature," *Scientific American*, May 1963, p. 45.

[7]J. A. Wheeler, C. W. Misner, and K. S. Thorne, *Gravitation* (San Francisco: W. H. Freeman, 1973), p. 1212.

expensive high-energy equipment. After the tour, the director approached her, and said "This equipment, then, is what we are using to unlock the secrets of nature." To this Mrs. Einstein replied, in a rather puzzled fashion, "But that is what my husband does too, unlocks the secrets of nature. But he does it with a pencil on the back of an envelope!"

In my view neither empiricism nor rationalism offers a completely adequate position for science. I am led to this view especially by consideration of the possibility that some of the various sciences are in principle different from one another, and thus might require different epistemologies. But we will hold off on discussion of this point until later. However, it at least should be clear from the present discussion that epistemological questions can and in fact *must* be raised within science. The question "How can we achieve scientific knowledge about the real objects of this world?" is a live and abiding issue for the scientist.

LOGIC

Logic is the set of answers to the philosophical question "Is the pattern of reasoning under consideration a correct one?" There are many aspects of patterns of reasoning which are subject to the scrutiny of logicians. For example, they can investigate whether or not the reasoning committed any fallacies, or whether it is valid, or sound, or whether it is inductive or deductive, and so on. The logician can be interested in *any* specimen of reasoning, no matter whether it occurs in a legal context, or an economic context, or in the context of campaigning for an election, or in the context of a scientific argument.

Logicians have been particularly interested in the reasoning of scientists ever since Aristotle first proposed a system of formal logic. With that proposal, it became clear that scientific procedure raises some important problems, and that those problems must be solved as part of the process of answering the epistemological question "What counts as scientific knowledge?" In Chapter 2 I will lay out some of the details necessary for a precise application of logical concepts in the theory of science. But first I will describe just one general example of the type of philosophic problem at issue. This way, you can at least become familiar with the flavor of logical problems before we get down to precise details in the next chapter. The case I am going to describe is a fascinating one, and one which can make nearly any scientist or science teacher just a tiny bit nervous.

Since the middle of the last century, it has been generally believed that nature conserves energy on a universal scale (indeed, the roots of this belief go all the way back to the time of Leibniz). For example, if two cars run into each other, the energy that each contains, the energy of mass in motion, does not disappear in the collision. Rather, it is converted into the stresses and strains of the bent metal, as well as the heat which has been lost to the surrounding environment. In a similar fashion, all energy reactions are understood to be nothing more than transformations of one kind of energy into another. Thus the chemical energy bound up in gasoline is converted to the energy of heat plus the force of the moving piston, when the gasoline is combusted in the cylinder of an engine. Similar conservation laws are true for other kinds of systems; for instance, it is generally held that mass is not lost in any kind of reaction, unless it be transformed into pure energy as described according to Einstein's famous equation, $e = mc^2$. The usual statement in contemporary science is, "Mass and energy are conserved in all physical interactions." The consequences of this statement for science are extremely broad in scope. One especially significant consequence is that mathematical equations have come to be a central means of our descriptions of observed physical occurrences. It is clearly the case that if we did not believe that mass and energy were conserved, it would not be so easy and natural to conceive of physical, biological, chemical, and other interactions in terms of equations. The French historian of science Emile Meyerson has traced the consequences of our belief in the conservation laws, and specifically shows how in many cases the data came to be arranged in equational form, long before it accurately reflected a mathematical equivalence.[8] His ultimate conclusion was that our use of equations is strongly linked to our belief in the conservation laws. This conclusion follows from the notion that equations describe balanced situations, that is, situations in which the interactions are merely rearrangements or transformations. Thus, use of an equation is congruent with the notion that the quantities are conserved from one side to the other of the formula.

Obviously, the conservation laws form a fundamental and essential part of science according to this analysis. Science as we know it simply would not exist if the conservation laws were not believed to be true. But as always, the philosopher must ask questions about procedure. In this case, we must ask the scientist a logical question: "What is the proof of the truth of the conservation laws? Since you believe that the conservation laws are true, then there must be some proof, some logical pattern of evidence which backs up your belief. Indicate the logic of your proof."

The typical response to this question will have been encountered by nearly all of us somewhere along the line in our education. In my own case, I can well remember the day in my ninth-grade science class when someone asked the teacher "But how do we know that mass is conserved

[8]Emile Meyerson, *Identity and Reality* (New York: Dover Publications, Inc., 1962).

in that chemical reaction?" The case in point had been a demonstration in which sulfuric acid was poured into a beaker of table sugar, a vigorous reaction ensued, and after all the heat and smoke subsided, a chunk of apparently pure carbon remained in the beaker. When asked the question about mass conservation, the teacher readily set up a simple, logical proof. He did the demonstration once again, but this time he did it on top of a scale. He weighed the reactants prior to the reaction, and then weighed them again after it was all over. The weight before was identical to the weight after; ergo, mass was conserved. As I soon found out, the demonstration was not entirely straight. For example, it was obvious that some steam vapor had escaped during the reaction, thus carrying off a small but observable part of the reaction's mass. Clearly, if the instructor had borrowed the high-quality balance used in the eleventh-grade chemistry classes, we would have noticed the weight difference between before and after. I am not accusing my teacher of deception, but . . . I am sure his motives were good ones, even if he did use a cheap balance.

But let us suppose that we are setting up a similar experiment to attempt proof of the mass-energy conservation laws. This time, however, we can stipulate that the experiment will be set up in such a way that absolutely no energy and no mass can escape from the experimental setup. For example, suppose that the reaction will take place in the box S, which is placed upon the balance scale M (see Fig. 1-2a). S, the experimental system, will be constantly measured by the measuring device M, throughout the reaction. M, by hypothesis, is as accurate as we can make it using contemporary techniques. What happens when we run the reaction? The prediction is that, if conservation is true, then, if S really is tight, and M really is accurate, the dial reading of M will not change during the reaction. And when we in fact run the experiment, presto! the dial on M does not change. Ergo, we have proved the conservation of mass-energy, right? Wrong. The proof contains a vicious logical error, one which vitiates the entire result of the experiment.

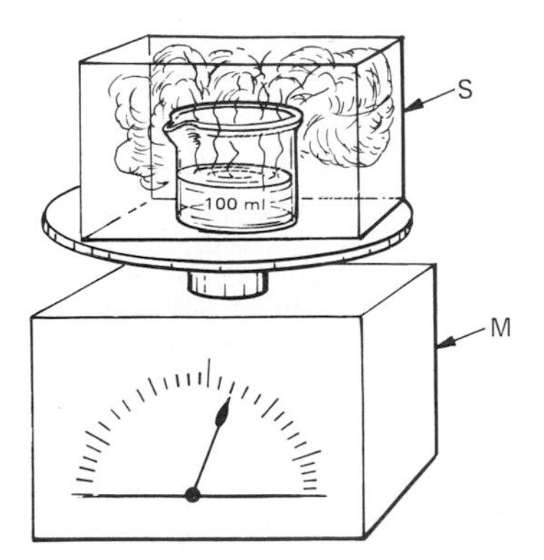

Figure 1-2a. Reacting system S, measured by scale system M.

Consider M: Just what *is* a balance device? There are many sorts, but most of them consist of some sort of mechanism which strains against the object which is to be measured. Thus, a coiled spring, or a twisted beam, is set up in such a way that it exerts its potential force against the system which is to be measured. The difference between the spring or beam potential force *before* it encounters the measured system and the spring or beam potential force *after* it encounters the measured system is what shows up as the dial reading. Thus, what the dial indicates is a difference between two forces—the before and after forces—within the measurement device itself. What I am going to say has probably already occurred to you, but I will say it anyway: the dial reading which we get from M can be taken to be accurate *only if* the law of conservation of mass-energy is true for M within itself *while we are using it to measure S.* To see this, just suppose for instance that the internal energy state of M decreased during the interaction. This would be like saying that the spring force potential weakened during the measurement. But if this happened, then if S lost energy during its reaction, the dial of M would still be stationary. Thus, in this example, although the experiment apparently proves conservation because the dial does not move, conservation in fact is not demonstrated. The point is that, in order to prove that the law of conservation of mass-energy is true for S, we must first *assume* (not *prove*) it to be true for the device M.

Of course we could observe whether or not M conserved energy-mass by enclosing it in another, larger, tight box S′, and putting S′ onto *another* measuring device M′. (See Fig. 1-2b.) If we carried out the reaction in S, and M′ stayed constant, energy-mass would be conserved in the system S′—and thus in M and S as well. But here is the kicker: The same problem would arise again, namely, S′ is proved to be energy-mass conservative only if the measuring device M′ can be taken to be energy-mass conservative. It is obvious that we could go on enlarging the measured system, going to S″ and S‴ and so on until the system included the whole universe, S^u. The question "Is the universe energy-mass conservative?" cannot be answered unless there is some measuring device which can measure the whole universe—and of course, even in that case, we would still have to *assume,* not *prove,* that the universal M^u conserved mass-energy, which would require a M^{u+1}, and so on.

There is a logical pattern here which contains a fallacy. The fallacy is called "begging the question," and it occurs whenever an argument, or proof, is offered as an answer to a demand, but the proof itself assumes, and does not prove, the very answer to the demand. Thus, in the case above, the question "Is energy conserved?" is apparently answered by demonstrating that energy is conserved in the system S. Thus, if energy is conserved in S, then it is conserved in the universe. However, as our

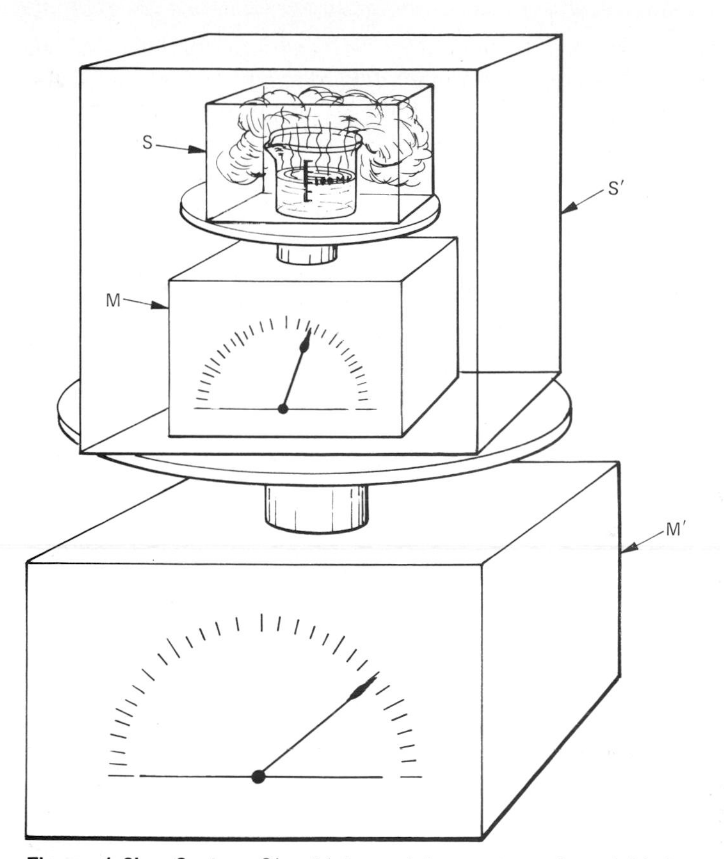

Figure 1-2b. System S′, which contains systems S and M, is measured by scale system M′.

closer examination showed, energy-mass is proved to be conserved in S only if it is *assumed* to be conserved in measuring device M. And so on. Thus, any experimental proof of the conservation laws depends upon first assuming the truth of the conservation laws. Consequently, any proof of the conservation laws begs the question whether or not the conservation laws are true.

There is no way around the logical error we are discussing. It simply cannot be proved via experiment that nature is in fact conservative of mass and energy. Although this might very well be somewhat disconcerting to you, it probably should not be. Rather, it should alert you to the point I made earlier, namely, that science is a lot different from what it is usually taught to be. But this difference does not mean that science is useless, or false, or hypocritical, etc. It simply means that it is different than you have been taught. For example, although valid logical proofs of the conservation laws cannot be given via experiment, other kinds of arguments can be given. One, for example, is the simple-minded one that, "Science depends upon the conservation law; science works; therefore, the conservation laws are accurate." I do not think too much of this particular argument. Better arguments can be given; we will see some of them later, in the discussion of Lavoisier in Chapter 5.

At this point, I hope that you have a somewhat better feel for what a logical issue in science is. Logical issues concern the methods and patterns which are used in reaching scientific knowledge. In fact, as I shall indicate in Chapter 2, some logical patterns of reasoning can afford absolute guarantees about our reasoning. This point alone should suffice to show the significance of logical elements in the theory of science.

SUGGESTIONS FOR FURTHER READING

The *Encyclopedia of Philosophy*, Paul Edwards (ed.) (New York: Macmillan, 1972), is an especially useful source for general philosophical information. For example, the articles under "Empiricism" and "Rationalism" are excellent introductions to these subjects. It is also possible to find articles under philosopher's names, for example Leibniz and Bacon. Very general articles, such as "Logic" and "Epistemology," are also of some use.

One specific topic in Chapter 1 has been discussed many times, and may be of some interest to you. This concerns the debate over the logical and metaphysical problems in the atoms versus forces case. Probably the best sustained discussion of this issue is the relatively unknown but excellent essay by the Cincinnati jurist J. B. Stallo, *The Concepts and Theories of Modern Physics* (Cambridge: Harvard University Press, 1960). A more recent account of the controversy can be found in Rom Harré, *The Principles of Scientific Thinking* (Chicago: University of Chicago Press, 1970), especially chaps. 10 and 11.

Chapter 2

A Precise Language

THE NEED FOR A PRECISE LANGUAGE

Every field of study has its own special language, with its own special words, meanings, and concepts. The reason for this is not too difficult to find: Each discipline looks very closely and precisely at a restricted part of the universe. This sort of close inspection and precision requires a clarity of linguistic function far beyond that found in everyday language. Thus we are forced into technical, manufactured languages for each discipline. Of course, as modern examples have shown us—bureaucratese, pentagonese, educationalese are only some of the *extreme* cases—this development of special languages can be overdone to the point where the average citizen cannot understand what is being said; and this is risky in a democracy. But even given these lamentable excesses, there still remains a real need for technical languages in the various special fields of study.

The theory of science is no different from other disciplines in this regard, and consequently a technical language has developed. For the most part, the fundamental elements of this technical language have been borrowed from modern formal logic. There are several reasons for this borrowing. In the first place, a view held by many investigators in the theory of science includes the belief that science, by nature, consists of concepts and pieces of reasoning which can accurately be modeled by formal logic. Although this view no longer completely predominates the field, it has had enormous influence upon the historical development of the theory of science. Second, the reasoning processes of science (and scientists) have been a prime focus of scrutiny for over 2000 years. This intense and continuing analysis of the methods of scientific reasoning has typically been motivated by a belief either that scientific reasoning is different from other types of reasoning, or that scientific reasoning is an almost ideal example of all human reasoning. In either case, what results from the motivation is an attempt to describe the "logic" of science. A third and final reason that terms and concepts borrowed from logic form the major part of the technical language used to describe the history and philosophy of science is that such terms are very useful in the construction of further new terms descriptive of science. Hence, for all three of these reasons, we need to become familiar with some fundamental concepts of modern logic. But before introducing the logical material, I want to say a few things about contemporary logic.

Contemporary logic is chiefly the work of mathematicians and philosophers who worked during the last parts of the nineteenth century and the early parts of the twentieth century. It is a curious discipline. The initial concepts of formal logic were first discussed by Aristotle in the fourth century B.C. There had been almost no hints of formal logic in the works of earlier philosophers, but completely on his own, he produced a system of logic which has endured, with some changes, until now. Following Aristotle, further important developments occurred during Greco-Roman and medieval times. But the most significant initial insight into the possibility of modern logic was proclaimed by Gottfried Leibniz, whose acquaintance you made in Chapter 1. Leibniz discovered that symbolic logic held the potential to be the foundation, the ultimate connection, between what had been previously thought to be two vastly different modes of human cognition, namely, mathematics and language. Leibniz himself was not able to move very far into the field, beyond a few extremely suggestive first beginnings.[1] But he was able to show the definite possibility of a strong yet previously unsuspected connection between grammar and algebra, between linguistics and mathematics. It is this striking blend of two apparently very different fields which most

[1]G. G. Leibniz, "The Method of Mathematics," in I. M. Copi and J. A. Gould (eds.), *Readings in Logic* (New York: Macmillan Co., 1972), p. 185. This collection of readings contains quite a few very nice short essays about logic; I recommend it highly if you are interested in this sort of thing.

likely has motivated modern philosophers and mathematicians to develop the many concepts of contemporary logic. In the remaining discussion of this present section I will introduce you to some—although by no means to *all*—of the concepts of modern symbolic logic. However, before getting into the details, I must warn you of something. Logicians tend to write very tersely. That is, they compress into one sentence enough material for two or three more leisurely sentences. Although I am not a logician, I must admit to being a bit guilty of this sin in the following discussion. However, if you read, and reread, the discussion carefully, the main concepts should be clear enough to let you proceed in a worthwhile manner.

SYMBOLIC LOGIC[2]

Simple Statements

1 English (like all other natural languages—German, Latin, Swahili, etc.) contains many *simple statements* of fact. For example, "It is raining" and "The streets are wet" are simple statements. "It is raining and the streets are wet" is *not* a simple statement.

2 Simple statements are either true or false, but not both. The *truth value* of a statement is T or F; that is, if the statement is true, then its *truth value* is T. If it is false, its *truth value* is F.

3 We may abbreviate, or *symbolize*, a simple statement by assigning to the statement a corresponding capital letter. For example, we may let "R" stand for "It is raining"; and "W" stand for "The streets are wet." Thus, we say that "R" is the *symbolization* of "It is raining." In other terms, "R" is a *symbolic* statement, and "It is raining" is an *English* statement.[3]

4 The *truth value* of a *symbolization* of a statement is the same as that of the statement itself. Thus, if "The streets are wet" is T, then "W" is also T.

Compound Statements

1 Simple statements of English may be connected to produce compound statements. For example, "It is raining and the streets are wet" is an English compound statement produced by connecting "It is raining" and "The streets are wet" through the use of the English connective

[2]My treatment of the formal system closely follows that of Irving M. Copi, *Introduction to Logic* (New York: Macmillan Co., 1972), chaps. 8–10. However, my choice of the arrow ('→') to represent the conditional comes from my own first introduction to modern logic in the work of D. Kalish and R. Montague, *Logic* (New York: Harcourt, Brace & World, Inc., 1964).

[3]All statements, whether in English or symbols, will be identified by double quotation marks. All special logical signs, or English logical names, will be identified by single quotation marks when they do not occur within statements.

'and.' Note that the compound statement has a truth value. That is, the compound statement "It is raining and the streets are wet" is T only if "It is raining" is T and "The streets are wet" is T. If either of the simple statements is F, then the compound statement is also F. Thus, if it is not raining, then "It is raining and the streets are wet" is F.

2 Symbolic statements, just like English statements, may be connected to produce compound statements. For example, if we invent a symbol to stand for 'and,' we can produce a symbolization of "It is raining and the streets are wet." Let "R" and "W" symbolize the simple parts of this symbolization. Now let '.' (called 'dot') symbolize 'and.' Thus, the compound statement "It is raining and the streets are wet" may be completely symbolized by: "(R.W)."

3 Note that the truth value of the whole statement "(R.W)" depends upon the truth value of "R" and the truth value of "W," just as it does in English. Thus, if "R" is T, and "W" is T, then "(R.W)" is T. In any other case, "(R.W)" is F.

4 We shall be interested in only five different types of compound symbolic statements:

a 'Not' statements: Simple statements may be denied. For example, we might say "It is not the case that it is raining" (or more simply, "It is not raining"). 'Not' is symbolized by '~,' which is called 'tilde.' The truth value of a 'not' statement depends primarily upon the truth value of the simple statement from which it is formed. For example, suppose it is T that "It is raining." If this is true, then we must say that it is false that "It is not the case that it is raining." Thus, if "R" is T, then "~R" is F. But if "R" is F, then "~R" is T.

b 'And' statements (also called 'conjunction' statements): These were discussed above.

c 'Or' statements (also called 'disjunction' statements): These statements are roughly similar to English statements which are formed by connecting simple statements with 'or.' For example, "It is raining or it is snowing." If we let "R" stand for "It is raining," and "S" stand for "It is snowing," and 'v' (called 'wedge') stand for 'or,' then the symbolization of this compound statement is "(RvS)." By definition, the truth value of a disjunction is F only when all the simple statements are F. Thus, "(RvS)" has the truth value F only when "R" is F and "S" is F. In all other cases, "(RvS)" is T.[4]

d 'If, then' statements (also called 'conditionals'): These statements

[4]Note the case where *both* "R" *and* "S" are true. Then consider the English statement "It is raining or it is snowing" when the weather is doing both. Is the English sentence T, or is it F? Some people would say T while others would say F. This result indicates that English is ambiguous about 'or.' Lawyers recognize this fact about English. Thus, contracts always specify "and/or," or "one or the other, but not both." Logicians have settled upon the "and/or" meaning of 'v.' Thus, 'v' is T when both sides of it are T. Although this might seem a bit arbitrary, it works out well because compensation for 'v' has been made in other parts of the symbolic system. Thus, any instance of 'or' in English will be symbolized by 'v' without worrying about the correctness of such a usage in the particular situation.

are roughly similar to English compound statements formed by connecting two simple statements with 'if, then'; for example, "If it is raining, then the streets are wet." If we let "R" and "W" be the same as above, and let '→' (called 'arrow') stand for 'if, then,' then the complex statement "(R→W)" symbolizes the English statement "If it is raining, then the streets are wet." The truth value of "(R→W)" is F only when "R" is T and "W" is F. Otherwise, "(R→W)" is T.

e 'If and only if' statements (also called 'biconditional' statements, or 'definition' statements): These statements are not terribly common in English, although they are relatively common in mathematics and science. One example might be "He is a bachelor if and only if he is an unmarried man." Note that this looks very much like a definition of the terms "bachelor" and "unmarried man." If we let "B" stand for "He is a bachelor" and "U" stand for "He is an unmarried man," then the symbolization of the biconditional compound statement is given by "(B≡U)." The truth value of the biconditional is T only when the truth values of both simple statements are identical. For example, if it is true that "(B≡U)," then if it is true that he is a bachelor, then it is also true that he is an unmarried man. Moreover, if it is true that "(B≡U)," then, if it is false that he is a bachelor, then it is also false that he is an unmarried man. (I know this sounds strange, but think it over several times.)

Truth Tables

1 There is a means for defining all the possible ways in which the truth values of statements may vary. Think back about '.'. Suppose we are studying two simple statements connected by 'dot,' and we can let these two simple statements sequentially adopt all the possible variety of individual truth values. Although you might think that would be a large number of combinations of truth values, in fact there are only four different cases:

Case 1 Both simple statements may be T.
Case 2 The first statement may be T and the second F.
Case 3 The first statement may be F and the second T.
Case 4 Both statements may be F.

These four cases may be symbolized: Suppose that there are two simple statements "R" and "W" (as above):

	"R"	"W"
Case 1	T	T
Case 2	T	F
Case 3	F	T
Case 4	F	F

(The four horizontal lines, corresponding to the cases, are called 'lines.' The two vertical columns, corresponding to the statements, are called 'guide columns.')

These four cases correspond to *all* the possible ways in which the truth values of "R" and "W" may vary. We may now *define* the truth values for '.' by simply adding a third column (the '*definition* column') to the right of the two guide columns. For example:

	"R"	"W"	"(R.W)"
Line 1	T	T	T
Line 2	T	F	F
Line 3	F	T	F
Line 4	F	F	F

This definition says symbolically that a compound dot-statement is true *only when* both its simple statements are T. Otherwise, the dot-statement is F.

2 In just the same way, each of the other four connectives may be defined. The simplest one is '~.' Think about the statement "R," which symbolizes the English sentence "It is raining." "R" may have only two possible truth values: Either it is T, or it is F. Thus:

	"R"
Line 1	T
Line 2	F

(Note that there is only one guide column.)

We may now define the compound statement "~R", by adding the definition column to the right of the guide column.

		Definition of '~'
	"R"	"~R"
Line 1	T	F
Line 2	F	T

3 Similarly, each other symbolic connective may be defined. (Note that the guide columns for all the truth tables are identical.)

Definition of 'v'			Definition of '→'[5]			Definition of '≡'		
"R"	"W"	"(RvW)"	"R"	"W"	"(R → W)"	"R"	"W"	"(R ≡ W)"
T	T	T	T	T	T	T	T	T
T	F	T	T	F	F	T	F	F
F	T	T	F	T	T	F	T	F
F	F	F	F	F	T	F	F	T

4 The overall notion of a symbolic statement may now be defined. This definition gives the grammatical rules for symbolic statements, and functions as a basis for deciding whether or not a given pattern of symbols is an acceptable pattern which qualifies as a symbolic

[5]Even cursory inspection shows that '→' is a very peculiar connective indeed. Some of its peculiarities are discussed at the end of the chapter.

sentence. English has rules which carry out a similar function. Let us look first at an English example. Consider the two following patterns of English symbols:

a Saturday is green.

b down the hill tumbled Jill

The question is, "Is either pattern a or pattern b a sentence of English?" The answer is, "No, neither pattern a nor pattern b is a sentence of English." But the reasons why each fails to be a sentence of English are very different from one another. The first pattern of symbols fails because there is a conflict between the underlying concepts or meanings of the words themselves. Basically, the subject concept 'Saturday'—a day of the week—cannot be linked with the predicate concept 'green,' simply because days are not colored. If the subject concept instead had been 'tree' or 'dress' or even 'Charlie's seasick face,' then a would have qualified as a sentence.

The second pattern, b, is also not a sentence, but for a very different reason. The concepts of b are quite all right, they square well with one another. Thus, b does not fail because its concepts do not mesh. Rather, b fails because it is drawn improperly, that is, the geometrical pattern of the ink on the page is inadequate for English. All English sentences, in order to qualify as *sentences*, must begin with a capital and end with an appropriate punctuation mark, either "!" or "?" or ".". But b neither begins with a capital nor ends with an appropriate punctuation mark. Thus, b is not a sentence of English. Logicians use the term 'well-formed pattern' to refer to sets of symbols which qualify according to the grammatic rules. And b is *not* a well-formed English pattern since it does not satisfy the rules of English.

The well-formed-pattern rules of the symbolic system are very simple, although, as I shall immediately point out, they have some complicated implications. The rules use a couple of capital letters which have been agreed upon by logicians to have a unique function. The symbolic abbreviations 'P' and 'Q' have been selected to function as "variables,' that is, as symbols which represent other symbols, just as, in algebra, x can stand for 4 or 1001, or any number at all. In algebra, in fact, x simply *means* "some particular number." Along these same lines, 'P' just means 'some particular statement symbol' and so does 'Q.' Let me now give the rules.

Rules for Well-formed Patterns of Symbols[6]

Rule 1 'P' is a well-formed pattern, and thus is a symbolic statement.

Rule 2 '∼(P)' is a well-formed pattern, and thus is a symbolic statement.

Rule 3 '(P.Q)' is a well-formed pattern, and thus is a symbolic statement.

[6]My colleague Bruce Bubacz helped me a great deal in getting these rules put in properly, and in general he helped me quite a bit with this whole chapter, which is much appreciated.

Rule 4 '(PvQ)' is a well-formed pattern, and thus is a symbolic statement.
Rule 5 '(P→Q)' is a well-formed pattern, and thus is a symbolic statement.
Rule 6 '(P≡Q)' is a well-formed pattern, and thus is a symbolic statement.
Rule 7 There are no other well-formed patterns.

These rules are quite straightforward, and simply delineate concepts we have already discussed. For example, "I~' is not a well-formed statement according to these rules, and neither is "(P.Q.R)," although "[(P.Q).R]" is OK. However, just as with all other formal definitions, there is more here than meets the eye. Some sneaky ramifications come out upon further study. Remember that 'P' can stand for *any* other sentence. Ditto for 'Q.' Watch what can happen because of this. Suppose we set up the following abbreviations:

"R": "It is raining."
"W": "The streets are wet."

According to Rule 1, "R" is a symbolic statement, and so is "W." But this means that "(R→W)" is also a well-formed pattern, according to Rule 5. But, if "(R→W)" is a symbolic statement, then we can use 'P' to stand for it, since 'P' can be used to stand for *any* symbolic statement. *Now* we can begin to construct all sorts of sentences, even ones of very great complexity. Go back to "(R→W)." Is this pattern well-formed? Yes, according to rule 5. Can 'P' be used to represent "(R→W)"? Yes, because 'P' can represent *any* well-formed pattern. Thus, 'P' represents "(R→W)." A very complex question now gets asked: Is "~(R→W)" a well-formed pattern? Our answer must be "Yes." Why? Because 'P' stands for "(R→W)" and "~(P)" is well-formed according to rule 2. Thus, "~(R→W)" is a well-formed symbolic statement.[7]
Using this sort of procedure we can construct very long and complex symbolic sentences, and decide whether or not they are acceptable. Moreover, and this is a fascinating aspect of the whole process, these very long but acceptable symbolic sentences make complete sense when they are translated back into English. Let me give one final example to show how this procedure has these effects. Consider the symbolic pattern "(Pv~P)." This sentence translates into literal English as "Some particular sentence is true, or it is not the case that this particular sentence is true." In less literal but more stylish English, this means, "A sentence is either true, or it is false." Clearly,

[7]Rule 2 is usually modified by logicians in the cases where 'P' is a simple statement. That is, where 'P' stands for a simple statement, then '~P' may be written instead of "~(P)." This saves some ink and, actually, is a bit easier to perceive.

this interpretation of the symbols is easy enough to understand. But now watch this: Suppose we let 'P' stand for "(R→W)." "(Pv~P)" now becomes "(R→W)v~(R→W)." According to our rules, this last creation qualifies as a well-formed pattern, and is therefore a legitimate symbolic sentence. Does this same conclusion hold true in English? Surprisingly enough, it does. Thus, we literally render the complex symbolic sentence "(R→W)v~(R→W)" as "The particular sentence 'If it is raining, then the streets are wet' is true, or it is not the case that the particular sentence 'If it is raining, then the streets are wet' is true." More stylishly, we can say, "The sentence 'If it is raining, then the streets are wet' is either true, or it is false." I think you should now be able to see how the symbolic system builds and, moreover, that if you follow the simple rules of the system, you can move back into agreeable English with ease. Thus, even a very complex sentence such as "[(Rv~R).~(R.~R)]" makes sense when translated into English.

At this point, let me now go back and redraw the truth-table definitions given above in number 3. The new drawing will take advantage of the grammar rules and of the fact that 'P' and 'Q' are variables. Because of this latter fact, when I define '(P.Q),' I will be defining the connective '.' for *all* sentences, since the variables being used mean that '(P.Q)' translates as "some particular sentence *and* some particular sentence." This is a much more universal way to define '.' than I used above, since those definitions are limited to the sentences "R" and "W" as drawn. The truth-table definitions below apply to *all* sentences which can be represented by 'P' and 'Q' and the connectives.

'P'	'Q'	'(P.Q)'
T	T	T
T	F	F
F	T	F
F	F	F

'P'	'Q'	'(PvQ)'
T	T	T
T	F	T
F	T	T
F	F	F

'P'	'Q'	'(P → Q)'
T	T	T
T	F	F
F	T	T
F	F	T

'P'	'Q'	'(P ≡ Q)'
T	T	T
T	F	F
F	T	F
F	F	T

Validity

1 The technical notion of validity may be defined rather simply: A pattern of reasoning is valid if and only if it is impossible to lose truth during the process of reasoning in accordance with the pattern. (That is, if we start out with true statements, a *valid* reasoning process absolutely guarantees that we shall end up with a true conclusion.) What this means is the following. Reasoning, especially in science and

mathematics, proceeds in a step-by-step fashion, starting from ‘premises,’ or ‘data,’ or ‘definitions,’ or ‘axioms,’ etc. and ending up with a ‘conclusion.’ (The precise meanings of these terms will be made clear shortly.) What we want to guarantee is that no mistakes are made during the actual step-by-step procedures. This comes down to saying that, if we start out by *assuming the premise to be true,* then valid reasoning guarantees that the conclusion will also be true. Note this well: When we are considering whether a procedure is valid or not, we do not say that the premises are as a matter of fact true. Rather, we say that *if* the premises are true, and the reasoning is valid, *then* the conclusion is true. The question about the *truth* of the premises is different from the question of the validity of the pattern. We will discuss the truth question later on.

2 There are a number of patterns of reasoning which can guarantee that conclusions are reached with validity. These patterns are called *elementary valid arguments.* Each of these arguments can be checked against a truth table to see whether or not true premises can lead to a false conclusion. If there is no case on the truth table where true premises lead to a false conclusion, then the pattern of reasoning is said to be valid. There are only three argument forms of high interest for our scientific reasoning. These are diagrammed below. Each pattern is called a *specimen pattern of an elementary valid argument*:

Modus ponens (MP)[8]		*Modus tollens* (MT)		Disjunctive syllogism (DS)	
Premise 1	‘$(P \rightarrow Q)$’	Premise 1	‘$(P \rightarrow Q)$’	Premise 1	‘$(P \vee Q)$’
Premise 2	‘P’	Premise 2	‘$\sim Q$’	Premise 2	‘$\sim P$’
Conclusion	/ ∴‘Q’	Conclusion	/ ∴‘$\sim P$’	Conclusion	/ ∴‘Q’

The special argument patterns which are diagrammed here are highly simplified. However, with fairly straightforward modifications, we may use them to represent quite precisely a model of many examples of reasoning in science. But first, let us briefly analyze these argument forms.

3 MP says the following: If we assume that a conditional statement is true, and we also assume that the first simple statement (‘P’) in the conditional is true, then we may validly conclude that the second simple statement (‘Q’) in the conditional is true. Thus, if MP is valid, and premise 1 and premise 2 are true, then the conclusion *must* be true. For example, think back to our statement “If it rains, then the streets are wet,” “$(R \rightarrow W)$.” Let us assume that “$(R \rightarrow W)$” is true. [Note that we now only *assume* it to be true. But we might go about

[8]The names *modus ponens* and *modus tollens* come from the Latin. They mean, respectively, affirmative mode and denial mode. In classical logic, each was a part of what was called “hypothetical logic.” Thus, MP is the affirmative mode of hypothetical logic, and MT is the denial mode of hypothetical logic.

proving this, rather than merely assuming it. For example, scientific observation might well have led us to conclude that, in all cases we have investigated, if it rains, the streets are wet. Thus, on the basis of our observations, we might take "(R→W)" to be a generally true-in-fact conditional.] Let us also assume that "R" is true, i.e., that it really is raining. What follows from these assumptions? It would seem that it follows that the streets are wet.

Consider another example. Let us assume that the following conditonal is true:

> "If this is a cell from a human male ("M"), then it will contain at least one Y chromosome ("Y")."

Suppose further that it is true that:

> "This cell is from a human male ("M")."

It would seem to follow as a conclusion that:

> "Therefore, it will contain at least one Y chromosome ("Y")."

We can represent this step-by-step reasoning by the following symbolic pattern:

P.1 "(M→Y)"
P.2 "M"
C. /∴"Y"

It should be noted that the argument pattern *modus ponens* represented here begins to look a little bit like the structure one would find in a scientific prediction. This suspicion would not be unreasonable, since many philosophers and scientists have claimed that predictions deduced from theories have the structure of an MP argument pattern. We will discuss this at greater length later. Note that the definition of MP takes advantage of the fact that 'P' and 'Q' are variables. Thus, both stand for '*any* particular symbolic statement.' Suppose as an example that we let 'P' stand for "A," and 'Q' stand for "~B." These abbreviations are acceptable since both "A" and "~B" are well formed according to the rules. Thus, given that the MP specimen pattern is:

P.1 '(P→Q)'
P.2 'P'
C. /∴'Q'

then, using the "A" and "~B" identically to the way that "P" and "Q" are used in the specimen pattern, we get

P.1 "(A→~B)"
P.2 "A"
C. /∴"~B"

This is perfectly correct. You can see from this that the MP pattern might have a great degree of symbolic complexity, yet still have the same simple, valid pattern. For example, "(A.~B)" can be represented by the specimen pattern's 'P,' and "(C.D)" can be represented by the specimen pattern's 'Q.' A valid MP interpretation according to this then would be:

P.1 "[(A.~B)→(C.D)]"
P.2 "(A.~B)"
C. /∴"(C.D)"

The specimen patterns can be used in such a fashion for each of the three arguments. The ability to make the specimen patterns fit greatly more complex arguments assures us that the standard of validity can be applied in an amazingly large number of arguments. The benefits of this capability for scientific reasoning should be pretty obvious.
If you have some doubts about the validity of MP, you might wish to attempt to find a case in which you start from true premises (i.e., a true conditional, and a true first simple statement in the conditional) and conclude with a false conclusion (i.e., a false second simple statement in the conditional).

4 Suppose that we assume for now that MP illustrates the pattern of a correct scientific prediction. What happens if the prediction goes wrong, that is, suppose that the prediction turns out false? The argument pattern which has been claimed to illustrate this feature of scientific reasoning is MT. Consider the following case. It is well known that Lamarck's theory of evolution contained the general view that offspring would inherit the characteristics acquired by their parents. One particular example of this general premise would be the following true conditional:

"If Lamarck's theory of evolution is correct, then male offspring of a circumcised parent will inherit this acquired characteristic."

However, as thousands of years of observation have indicated:

"It is not the case that male offspring of a circumcised parent will inherit this characteristic."

This observation forces us to state that:

"Therefore, we must conclude that Lamarck's theory is not correct."[9]

We can symbolize this argument. First we set up the following scheme of symbolic abbreviation to express the simple statements of the MT argument.

"L": "Lamarck's theory is correct."
"M": "Male offspring of a circumcised parent will inherit this characteristic."

Then, using these symbols, the pattern of the MT reasoning would go like this:

P.1 "L→M"
P.2 "~M"
C. /∴"~L"

I would again suggest that, if you have doubts about the validity of MT, then you might attempt to come up with an example in which the first and second premises (i.e., 'P→Q' and '~Q') are both true, but the conclusion '(~P)' is false.

5 DS should be familiar to everyone. It is ordinarily known as the "process of elimination." Consider the following example:

"The trouble is either in the electrical system, or in the fuel system.
"However, I've checked it out and the trouble is not in the electrical system.
"Therefore, the trouble must be in the fuel system."

We can symbolize this argument as follows:

"E": "There is trouble in the electrical system."
"F": "There is trouble in the fuel system."

P.1 "(EvF)"
P.2 "~E"
C. /∴"F"

[9]This example reminds me of a cartoon we once had in our biology department. The first panel showed a pregnant mother, father, and four kids, all of whom had identical, huge, grotesque noses. The second panel showed them marching into an office marked "Plastic Surgeon." The third panel showed them all lined up before the office, smiling, with beautiful, petite noses. The final panel showed the whole family gathered around a crib containing a newborn baby. They were all frowning. The baby, of course, had a huge, grotesque nose.

It should be obvious to you, without testing, that this argument pattern can never lead from true premises to a false conclusion.[10]

Deduction

Deduction is a process of reasoning which consists of fairly long chains of logical inferences. Deductions start from certain particular *given* statements of belief, or fact, or definition, and so on. These given statements can be quite varied in terms of their origin. In mathematical deductions (such as geometric proofs), the given statements may be theorems (statements which have been previously deduced from axioms) or hypotheses (statements which the mathematician hopes to prove or disprove). Legal "deductions" or reasonings, although not formulated explicitly in logical language, contain given statements such as police reports, or statements about earlier legal precedents, or statements reporting observations of witnesses, etc. In scientific deductions, the given statements are theoretical axioms (the initial assumptions of theories); or fairly general principles of nature (such as the principle of the conservation of mass and energy); or particular observation reports (such as the report of the result of an experiment).

The vocabulary of the logician may be used to describe the elements of each of these examples. First, each starts out from given statements, which the logician calls *premises.* Second, each moves from the premises to an ultimate, sought-for statement called the *conclusion.* The status of the conclusion statement is different in each field. For instance, in geometry the conclusion of a deduction would be a new theorem. In science the conclusion of a deduction might be a prediction about a new potential observational event. In law, the conclusion ultimately would be the verdict "Guilty" or "Not guilty." However, in addition to premises and conclusion, deductions have elements which connect these two parts. The connection is provided by logical inferences, that is, movements of the mind from the given statements to new, ungiven statements. The inferences occur in a particular, special, step-by-step fashion, a fashion which provides a complete guarantee of logical validity: Each inferential step in a deduction must correspond to the specimen pattern of an elementary valid argument. It should be obvious what results from this procedure. If each and every inferential step in a deduction is backed up by a specimen pattern of an elementary valid argument, then each and every one of the individual steps is valid. However, if each and every step in a deduction is valid, then it must be the case that the deduction as a whole

[10] This pattern of argument is set up to compensate for the fact that we have elected to symbolize all occurrences of English "or" by 'v.' Note that the second premise is a denial of one side of the 'v.' The first line of the truth table, where both sides are T, is automatically excluded. Thus, use of 'v' can never lead to invalid reasoning because of missymbolizing of English "or."

is valid. This overall, total validity has the ultimate and significant consequence that, no matter how many premises a valid deduction has, and no matter how long the deduction is, the conclusion must be true if the premises are true. Scientific reasoning takes particular advantage of this feature of deduction. Thus we find that scientific deductions are used to predict experimental conclusions on the basis of a set of premises which contain major elements of a theory. The general rule is that, if it can be shown that the predicted conclusion follows validly from a theory, then, if the theory is true, the predicted conclusion must be true. Thus, any disagreement between the validly deduced conclusion and the actual observed results would seem to indicate that the theory needs further work.

Let us consider the following example of how a deduction may be carried out in modern physics. The example is an obvious oversimplification of content, but its form is an accurate model. First, we must state the given premises which set up the deduction.

P.1 "Either Newton's theory is correct, or Einstein's theory is correct."
P.2 "If Newton's theory is correct, then there is no absolute top observable velocity in the universe."
P.3 "If Einstein's theory is correct, then the speed of light is the highest observable velocity in the universe."
P.4 "If the speed of light is the highest observable velocity in the universe, then the universe has an absolute top observable velocity."
P.5 "The speed of light is the highest observable velocity in the universe."

On the basis of these premises, the question we want now to answer is: Which theory is correct? Your intuition tells you already that "Einstein's theory is correct" is true on the basis of these premises. But what precise steps did your intuition take, and is your intuition guaranteed valid? To decide this question, let us see whether we can validly deduce "Einstein's theory is correct" from the given scientific premises. If we can, then we know that it is valid and true, assuming that the premises are true. To set up the deduction, we must first assign symbols to the premises:

"N": "Newton's theory is correct."
"E": "Einstein's theory is correct."
"A": "There is no absolute top observable velocity in the universe."
"L": "The speed of light is the highest observable velocity in the universe."

Now we list the premises symbolically. Note that each premise corresponds to exactly one English sentence from the example above.

P.1 "(NvE)"
P.2 "(N→A)"
P.3 "(E→L)"
P.4 "(L→~A)"
P.5 "(L)"

Now, using the three elementary valid patterns, we may construct a deduction. It is done as follows. We may compare sets of the premises to particular elementary valid argument patterns. If the sets of premises can be placed in a pattern identical to the pattern of the *specimen elementary valid argument*, then a conclusion identical to the one in the specimen may be reached. This conclusion is called an *intermediate conclusion.* Since the specimen pattern is known to be valid, and thus its conclusion is valid, the intermediate conclusion drawn from the premises by a pattern *identical* to the specimen means that we can plug the intermediate conclusion back into the premises and use another pattern to reach another intermediate conclusion. When we have done this a number of times, and the final result is the sought-for ultimate conclusion, then the deduction is over, and we have reached a valid conclusion.

To start, it should be obvious that we can connect P.4 and P.5 together, use MP, and conclude validly that "~A" is true. We illustrate this by noting below the line of the argument which deduction pattern has been used, and comparing it to the specimen pattern.

Line 1. Intermediate conclusion: "~A," because P.4, P.5, and MP.

	Deduction Pattern	*MP—Specimen Pattern*
P.4	"(L → ~ A)"	'(P → Q)'
P.5	"L"	'P'
Step 1	/∴ " ~ A"	/∴ 'Q'

MP allows us to conclude the second statement of a conditional, given the conditional and its first statement. Thus, "(~A)" has been validly concluded. However, once we have reached "(~A)" and know that it is a valid intermediate conclusion, we can then plug it into P.2, use MT, and conclude "(~N)." Thus:

Line 2. Intermediate conclusion: "(~N)" because line 1, P.2, MT.

Deduction Pattern		*MT—Specimen Pattern*
P.2	"(N $\rightarrow$ A)"	'(P $\rightarrow$ Q)'
Line 1	"(~ A)"	'(~ Q)'
Line 2	/∴ "(~ N)"	/∴ '(~ P)'

MT allows us to conclude the denial of the first statement of a conditional, given the conditional and the denial of its second statement. Thus, we have validly concluded "(~N)." We can now finally conclude "E," which is our sought-for ultimate conclusion. We do this on the basis of P.1, line 2, and DS. Thus:

Line 3. Ultimate conclusion: "E," because P.1, line 2, and DS.

Deduction Pattern		*DS—Specimen Pattern*
P.1	"(NvE)"	'(PvQ)'
Line 2	"(~ N)"	'(~ P)'
Line 3	/∴ "E"	/∴ 'Q'

This entire procedure may be cleaned up and presented in the fashion of a mathematical proof, as is done below. The statements on the left are either premises or the results of inferences from the premises. The notes on the right indicate the reason which justifies each left-hand statement. Reasons opposite inferences on lines 1 to 3 indicate what allowed each inference to be drawn from the lines immediately above it.

Question: On the basis of the premises given, can the statement "Einstein's theory is correct" be validly deduced from the premises?

Deduction		*Reasons*
P.1	"(NvE)"	Given
P.2	"(N $\rightarrow$ A)"	Given
P.3	"(E $\rightarrow$ L)"	Given
P.4	"(L $\rightarrow$ ~ A)"	Given
P.5	"L"	Given
Find:	"E"	Deduce, if possible, from the given premises.
Line 1.	"(~ A)"	"(~ A)" is inferred from P.4 + P.5 arranged in the MP pattern.
Line 2.	"(~ N)"	"(~ N)" is inferred from P.2 + line 1 arranged in the MT pattern.
Line 3.	"E"	"E" is inferred from P.1 + line 2 arranged in the DS pattern.

It must be noted what is achieved in this example. In its ideal form, this example shows one logical way to test between the Einstein theory and the Newton theory. According to this example, the logical test indicates that the Einstein theory is the correct one. That is, since the deduction is valid, the conclusion *must* be true, if the premises are true. Obviously, one big question remains: How did we obtain the original premises—that is, how did we get premises P.1 to P.5? According to the contemporary symbolic logical interpretation of scientific reasoning, these premises themselves have earlier been deduced from each of their respective theories. That is, if a scientist considered all the statements in Newton's theory, he could deduce premise P.2 from Newton's theory. On the other hand, if the scientist were to work carefully with the statements of Einstein's theory, he could logically deduce P.3 from that theory. (P.5, of course, is understood to be a statement of observational fact.) The ultimate conclusion, "E," is thus seen to be the result of a deduction from premises, which themselves are the result of either observation or previous deduction from other premises. It will be seen from this example that, according to this logical model, all the important reasoning in science consists in, first, setting up the premises of a theory and, second, using long chains of deductive reasoning to make predictions from the theory. A theory is considered well-tested—or well-confirmed—if the predictions deduced from the theory continue to be observed. A theory is considered to be doubtful when deductions from the theory do not turn out as predicted. One question remains: On this model of scientific reasoning, from whence come the important premises? Let us turn to that discussion.

Generalizations

It is generally believed that one of the chief functions of scientific research is to discover new truths about the physical universe and then to state these truths in the form of generalizations. Generalizations of this sort are often called "scientific laws," or "laws of nature," or some similar name using the term "law." Although there are some very crucial difficulties with this conception of science and its laws, I believe that the conception is useful enough for us to proceed into a discussion of it at this point, holding until later a discussion of the criticisms which have been brought against it. As we go along, you will find that our growing ability to precisely describe both the generalizations and the processes leading to them is of great help in understanding one of the central issues in the logic and epistemology of science.

The usual description of scientific generalizations is based upon the connectives which we have already discussed plus one additional logical element. The new element is entitled *quantification theory*. The term "quantification" refers to the fact that generalizations are in part different from other types of statements simply because they refer to quantities of things, rather than just to singular things. Quantification theory adds a vast degree of power and richness to our ability to symbolize sentences from natural language. It also makes much more precise the notion of validity as it applies to some very complex kinds of arguments. But, as with everything in this world, these gains are accompanied by a loss: Quantified sentences are more complicated than the simple symbolic sentences you have already seen. Moreover, as I shall now show, quantification theory requires that we adopt a special metaphysics, a special theory about how the world is constituted.

According to the metaphysics of quantification theory, the world is constituted by two sorts of things: first, individual objects, and second, the features or characteristics of these objects. The bottom line of this theory is that individuals themselves, in their ultimate reality, consist of the entire set of all their features. It is assumed that an individual is just identical to, i.e., is nothing more than, the simultaneous colocation of all its characteristics. Clearly this view (even in the dangerously oversimplified fashion here presented) has some puzzling consequences which we might want to investigate seriously prior to fully accepting the view. But full acceptance is not necessary for us now, since at this point we are interested solely in being able to describe in a simple fashion the symbolic translation and depiction of scientific generalizations. But it must be noted that, if at a later point, you or I ultimately reject the metaphysical view which is the foundation of quantification theory, we must *then* necessarily reject as well the depiction of scientific generalizations which is given in its terms. This will be discussed later on.

In any case, according to quantification theory, the universe consists of individual objects and their features, or, as logicians put it, "individuals and their properties." Quantification theory thus requires that there be symbols for these two sorts of entities. Properties are represented by capital letters: thus, 'T' might stand for 'is tall' or 'is a teacher' or 'is tight' and so on, depending on what the situation requires.[11] 'B' might stand for 'is bald' or 'is a billiard ball' or 'is blighted' and so on.

Small letters are used to name the individuals to whom the capital-letter properties belong. There are two groups of small letters, each of which has a different function. The first group consists of the small letters 'a' to 't.' This group functions like proper names, that is, each of these letters can be used to name a particular, known individual. For example, 's' might be used to name Socrates or Steve or Sheila, whatever is required. We can also name particular objects which do not have their

[11]The fact that the capital letters in quantification theory are not statements, but rather are *parts* of statements, is indicated by using only single quotation marks, instead of the double quotation marks seen earlier.

own individual names, but which can be referred to by description. For example, we can let 'r' stand for 'the rock in my left hand' or for 'that roach over there in the corner.'

The second and final class of names is a somewhat odd one. It includes the small letters from 'u' to 'z.' These letters are called 'variables' and they function like pronouns, or vague, general names. They are similar in function to variables in algebra and are even read in the same way. Thus the algebraic sentence "(x + 1) = 5" is translated, "A number plus the number one equals the number five." In logical language, the variable 'x' (or 'u' or 'v' or 'w' or 'y' or 'z') is translated 'an object' or even 'an x.' The need for variables will be crystal clear in just a moment.

When we put the new symbols together, we end up being able to make a large variety of new kinds of statements. "Ts," for instance, might stand for "Socrates is tall" or "Sheila is a teacher" or whatever you had decided the letters were to represent.[12] Similarly, "Br" might symbolize "The rock in my left hand is bald" or "That roach over there in the corner is blighted." Thus, statements containing capital letters and the small letters 'a' to 't' tell us the features of definite, known individuals. But when we shift into the variable symbols, e.g., capital letter plus small 'u' to 'z,' we note that the English gets much more vague. For example, 'Bx' might stand for 'Object is bald' (no particular object is named here, just 'object'), or as it is usually read 'x is bald,' where 'x' stands for absolutely nothing in particular. This latter reading is the one which I shall use. It should be getting clearer to you that this new system of symbolization allows us to analyze statements into much finer chunks than we were able to do with the earlier system. Let me now introduce the final complication added by quantification theory, namely, quantifiers themselves.

A *quantifier* is a special sort of symbol which is placed before a symbolic statement. Quantifiers are not variables but are symbols called logical constants. They tell us whether the symbolic statements they precede refer to an entire group, part of a group, or no members of a group.

Quantifiers are used in conjunction with the connectives which were introduced earlier. Thus, a complete symbolic statement in the quantifier logic consists of a quantifier followed by a compound symbolic statement. The compound symbolic statement itself consists of individual or variable names and property symbols, held together by the connectives. To make this clearer, let me work through a literal symbolization of an English sentence into the quantifier symbols. Take for an example the sentence

[12]Since "Ts," "Sheila is a teacher," and "Socrates is tall" are statements, they are identified by double quotation marks.

"All physical objects have mass." First, we introduce the symbol for 'all,' which is called the *universal quantifier* since it is understood to be universally true for all the objects in the group named by the symbolic statement. The universal quantifier is symbolized by placing a variable letter between parentheses, such as '(x)' or '(z),' and placing this symbol before the statement. Thus a partial literal symbolization of the English sentence "All physical objects have mass" would read:

"(x)(physical x's have mass)." (1)

Next we must translate "(physical x's have mass)," which we can do by defining 'P' to mean 'is physical' and 'M' to mean 'has mass.' Finally, we translate this into the conditional form, using the arrow. Thus, we ultimately end up with a symbolic statement:

"$(x)(Px \rightarrow Mx)$" (2)

The process of literal translation *from* symbols *into* English might make this method even clearer. Suppose we start out with (2) as above,

"$(x)(Px \rightarrow Mx)$" (2)

First, we replace the quantifier with 'all' (or, as I prefer, 'every') which leaves us with the partial translation

"(Every x)$(Px \rightarrow Mx)$" (3)

Next we can translate the symbolic statement *within* the parentheses. A literal reading of this portion gives us

"If x is physical, then x has mass." (4)

The complete literal translation then reads

"Every x, if x is physical, then x has mass." (5)

This literal reading sounds clumsy, and indeed, you might even think that it has little or nothing to do with the English sentence "All physical objects have mass." But think about the *logic* of the sentence "Every x, if x is physical, then x has mass." What it says is this: Suppose you take some object, any object in the whole universe (that is, just select some 'x' from the collection of everything that exists); if you verify that the object is a physical object, then you will be forced to conclude that the object *also* has mass. Thus, every physical 'x' is a massy 'x,' since "If x is physical, then it is massy." This reading is given further evidence by some of the variations which might form a continuum between the unstylish but literal rendering (5) of the symbolic statement, and the more stylish statement "All physical objects have mass." For example, we could read the symbolic sentence as "If any object in the whole universe is physical, then it has mass," which results from simply replacing 'x' by 'object.' We could also read it as "If anything is physical, then it has mass," or as "For everything, if it is true that it is physical then it is also true that it has mass," and so on. As you can see by the form of statements of this sort, these are what are generally taken to be scientific generalizations. On this analysis, then, laws of science are to be understood as universally quantified conditional statements.

The process by which generalizations are thought to be reached can be given precise form in the quantifier language. Let us look at an example which focuses upon a chemist, working with physical substances. The process is called *induction*, which, as you might remember from Chapter 1, is the logical procedure popularized by Francis Bacon. Suppose that a chemist is examining a particular chunk of some substance, and call it chunk 'a.' She subjects this substance to certain tests and discovers that it has an atomic weight of 12. She subjects it to further tests, and discovers that its valence is +4. She can then symbolize these facts by the following scheme:

P: has an atomic weight = 12.
Q: has a valence = +4.
a: the first chunk I tested.

Thus the following statement is true:

"(Pa.Qa)"

Suppose then that the scientist continues her tests, this time on chunk 'b.' She discovers "(Pb.Qb)" is true of this second chunk. These sorts of investigations go on, until, finally, the scientist concludes that "Whenever I examine a chunk of something which has an atomic mass of 12, then it will also have a valence of +4. Therefore, if anything has an atomic mass of 12, then it will have a valence of +4." This procedure is called *inductive generalization*, where the statement following the "Therefore" is itself the generalization. We can symbolize the generalization by the following set of symbols:

"(x)(Px→Qx)"

where this is read in English: "It is true of anything (any *x*) that: If x has an atomic mass of 12, then x will have a valence of +4."

You will note that this example very precisely divides the process of induction into two phases. Phase 1 consists of finding properties of objects, and correlating these properties. Thus, phase 1 consists of a series of repeated observations of properties of different individuals; the observations are described in terms of conjunctions of properties. This model of scientific observation, and replication of experiment, is very strongly believed by behavioral and social scientists, particularly in America. Stated very simply, this view holds that correlations between individual properties are what are observed by scientists and these are described in terms of conjunctions. (I believe this view is incorrect, but will withhold comment until later.)

Phase 2 consists of the step, or logical inference, from the set of conjunctions to the universally quantified conditional statement. This inference, this movement of the mind from a set of particular data to a general statement, is the special characteristic of induction. Thus, this inference is called an *inductive inference.*

My own preferred term for this inference is *inductive leap.* As you can see, my terminology suggests that I believe the inference to involve a certain amount of danger, since otherwise I would not call it a "leap." Most of my philosopher and logician colleagues are also a bit leery of the passage from a set of conjunctions to a universally quantified statement. One easy-to-see reason behind our caution is exemplified by the question: "How many data are needed to guarantee the truth of the generalization?" This question pointedly raises the fact that the set of conjunctions describes *only* and *solely* the data which have been accumulated. But the generalization is true, supposedly, about *all* 'x's,' even those which will become data only in *future* experiments. Consequently, the generalization claims that future events will be identical to past ones. How do the data guarantee this prediction? The theory of probability has come more and more into use as a device which will somewhat guarantee that the data are sufficient to warrant the generalization. But probability does not provide a *certain* guarantee; it just suggests that accepting the generalization is reasonable. The logical difficulties with the passage from conjunctions of observations to universal conditionals have come to be collected under the rubric "the problem of induction," which, since the time of David Hume, has been a classic problem for scientists of the strict empiricist persuasion.

To conclude this discussion, I might note just one significant feature of the logical situations presented here. As I noted earlier, the inferences which are made during deductions can be guaranteed absolutely since they can be matched to elementary valid arguments, which in turn can be matched to truth tables. Because of this, deductive inferences are certain. But as I have just noted, inductive inferences are not certain; they can only be probable. This difference in logical status between the two types of inferences can be seen to be a differentiating feature between rationalism and empiricism. Insofar as rationalism depends solely upon deductive logic, rationalistic logic can be taken to be certain. But insofar as empiricism depends solely upon inductive logic, empiricist logic cannot be taken to be certain. However, the case is not nearly so clear and clean as I make it here. As I pointed out in Chapter 1, there are few, if any, scientific positions which are purely one or the other, rationalist or empiricist. Moreover, as we shall later discover, there are always trade-offs between philosophies. Thus, if rationalism gains because its logic is certain, then it must lose against empiricism in some other area, for example, in metaphysical matters. Although I will not here attempt to define what I mean by "gain" or "lose," I will make my meaning quite clear later on down the line.

In any case, we now are in possession of the necessary logical tools to enable us to describe much more precisely the elements of the theory of science.

CONDITIONS, CONDITIONALITY, AND CAUSES: SOME PROBLEMS

The conditional arrow and its truth-table definition may seem rather odd to you. This oddness is especially noticeable on the last two lines of the truth table. In great part the apparent oddness of lines 3 and 4 comes about for two reasons: (1) conditionality is a strange aspect of the world, and (2) English, because it is a language which in part describes the world, reflects the strangeness of conditionality. Consider the following example. Suppose you are standing in your kitchen, trying to find a gas leak. A friend wanders in, and prepares to light a cigarette. You warn him, "Don't!" He asks, "Why?" You proceed to tell him about the gas leak and to say "If you light up, then the place blows up." The complex statement, the conditional "If you light up, then the place blows up" is clearly true in this case. But consider its simple component parts, "You light up" and "The place blows up." Both of these components, as a matter of fact, are false at the time you make the compound statement. That is, he has not yet lit up, and there is not an explosion. This situation corresponds to line 4 of the truth table for the arrow, where the components of the conditional are both false, but the compound statement is still true. What this case seems to indicate is that the conditional relation among gas leaks, matches being lit, and explosions is somewhat timeless, and even in a peculiar way independent of the present truth values of statements describing it. Some natural languages other than English, which make frequent use of the subjunctive mood, probably better reflect the rather odd situation involved in conditionality.

Although English has a subjunctive mood, fewer and fewer speakers are using it. For example, one *should* say, "If I were king, then I would declare today to be a holiday." "Were" and "would" are English subjunctive verbs. However, more often than not, today's speaker would substitute "was" for "were" and "I'd" for "I would." English thus appears, at least as far as the "was" is concerned, to be blurring the distinction between the indicative and subjunctive moods. Interestingly enough, the arrow connective rather effectively mirrors this blurring, and replaces the two distinct grammars with one single basic notion, that of conditionality itself. Thus, the conditional truth table matches nicely what should be described in the subjunctive. Consider one example of this. I have noted that, if I were king, then today would be a holiday. Certainly it is false that "I am king," and it is also false (as I write this) that "Today is a holiday." But the falsity of these two components does not in the least affect the truth of the conditional, since I hereby swear to you that, were I king, then today would be a holiday. The only way this statement could be false would be if I were lying to you—which I am not. Line 4 corresponds nicely to this subjunctive use, just as it did in the case of the purely conditional relation above among matches and gas leaks and explosions.[13]

In scientific practice conditionality plays a very important role. Sometimes, for example, we would want to say something like "If there were no air friction, then this chunk of lead would fall and accelerate at 32 ft/sec per second," as is always revealed by measurement. Even given these two false simple statements, however, the conditional compounded of them is true. A similar case can be seen in the E.P.A. mileage statements. Each type of car sold must have its relative mileage performance tested. The tests end up being quoted as something like "If you drive your car perfectly, then you will get 30 m.p.g.," which is perfectly true according to the tests. However, as we all know, the real questions are "Who drives perfectly?" and "Who gets 30 m.p.g.?" No one! I know I certainly do not. But this is completely irrelevant to the truth of the conditional relation expressed in the E.P.A. statement.

[13]Subjunctive truth relations are extremely tricky, and form a newly burgeoning area of logic. My colleague Henry Frankel told me one of the neatest examples of the trickiness of the subjunctive. First, you should know that the verb "should" is in the English subjunctive, just as is "would." Here's the example: Frankel was talking to a student about ethical issues raised during wartime. The student made a claim, which Frankel disputed. As a rebuttal, the student began a counterargument with the phrase "You should have been in the war in Vietnam. . . ." Frankel interrupted with the reply that it was false that he, Frankel, should have been in Vietnam. "Indeed" Frankel continued, "It is universally false that '*Anyone* should have been in the war in Vietnam'."

Another interesting occurrence of conditionality is very important for science. This occurrence is reflected on line 3 of the truth table, the line on which the first simple statement is false, but the second simple statement and the compound statement are both true. One example of this you have already briefly seen. Recall the example about Lamarck. In general, Lamarck's theory of biological inheritance and evolution has been discredited by contemporary biologists, i.e., they believe it to be false. But as Lamarck originally devised his theory, it could be used to explain why offspring tend to resemble their parents. Indeed, since this resemblance is a central fact of reproductive biology, *all* evolutionary theorists, and not only Lamarck, must attempt to explain it. Accordingly, the propositions of Lamarck's theory logically imply that offspring tend to resemble their parents. Because of this logical relation, the conditional "If Lamarck's theory is correct, then offspring tend to resemble their parents" is of course true. However, as I just noted, contemporary biologists generally believe that "Lamarck's theory is correct" is false. Thus, in this case, the first statement of the conditional is false, while the second statement, and the conditional as a whole, are both true. The truth values here thus correspond to line 3 of the definition of arrow.

It should be clear from my examples just how odd conditionality is. Moreover, both the symbolic language and the English language reflect this oddness. However, logicians *do* have a bit of an advantage in this situation. They can simply point to the arrow truth table and say "*This* is entirely and exactly what conditionality means; I thus define it." Although this move might seem as arbitrary as the similar move noted in the discussion of wedge, I must again note that compensations have been made in the symbolic system. In fact, the restriction on wedge, namely, that *all* occurrences of the English "or" are to be symbolized by 'v,' is related to the oddness of arrow, and vice versa. The point to note is that each of these seemingly arbitrary moves is compensated for by the system as a whole. Thus, in the final analysis, although various component parts of the symbolic system introduced here might not *accurately* reflect English, the system as a whole is an *adequate* model for English (and all other natural languages), since the question of the validity of any given piece of reasoning in either language is totally identical. Thus, no truth values get lost in the translation.

There is one final problem having to do with arrow. This problem, however, is not a logical problem, but rather has its roots in a psychological and philosophical attitude which many, perhaps even most, people manifest. The problem is this: 'If, then' conditionals and arrows are often read as if they were causal relations. That is, the logical relation of conditionality sometimes takes on features of the metaphysical relation of causality. In many philosophical systems, this blurring of the distinction between the two relations is not particularly vicious. However, the blurring is troublesome for science. Let me give you some examples of what I mean.

In the conceptual systems of many scientific philosophers, the question arises: "How is it that thought—mathematical thought in particular—corresponds so closely to the world?" An alternative phrasing of this question might be: "How much does the structure of language and thought reveal about the structure of the world?" These questions raise real problems simply because we have so many examples of correspondences among thought, language, and the world. Thus, the correspondence is real, and must be investigated in an attempt to discover its underlying cause.

Philosophers have answered these questions in a number of ways. Some empiricists, for example, have claimed that the structure of thought and language patterned itself after our experience. That is, they believe that the world stamped its structure upon our cognitive activities. Other empiricists, however, have claimed that there is no natural relation between thought and the world; rather, there is just a correspondence decreed by God.[14] An extreme rationalist alternative to these empiricist answers is oftentimes given, namely, that our thought in fact "creates" the world—at least in the fine details. This position is somewhat reflected by the American psychologist Benjamin Whorf's view that a culture's language in many ways shapes the world that it perceives.[15] On this account, then, the Eskimo culture lives in a world which reveals much more detail about snow than does any culture which does not share their language and world. But this is an extreme rationalist position.

The typical rationalist position is simply to postulate that there are close correspondences between thought and the world, and to take on the problem from there in an effort to see what can be made of the position. Benedict Spinoza, a contemporary of Leibniz, stated this view most succinctly when he claimed that the order and connection of ideas was identical with the order and connection of things.[16]

But no matter what their overall positions, all these assorted philosophies tend toward the view that there is a close relation between linguistic or logical structure and metaphysical structure. Thus, logical conditionality begins to take on a relation to aspects of causality. This is especially true with regard to the universal conditional, which raises distinct

[14]John Locke, *Essay Concerning Human Understanding*, III, ii, I; in T. V. Smith and M. G. Grene (eds.), *From Descartes to Locke* (Chicago: University of Chicago Press, 1940), p. 405.

[15]Benjamin Whorf, *Language, Thought, and Reality* (Cambridge: Massachusetts Institute of Technology Press, 1956).

[16]Benedict Spinoza, "On the Improvement of the Understanding," in Smith and Grene, op. cit., p. 247.

problems for scientific thought. If you will recall the discussion earlier about generalization, you will remember that I pointed out that one typical pattern of reasoning in science consisted of the logical movement from collections of conjunctions, or correlations, to an ultimate generalization. This movement can be defined in symbols as the movement "(Pa.Qa).(Pb.Qb) . . . (Pn.Qn). Therefore, (x)(Px→Qx)." Now this movement is perfectly all right in a logical sense, since there are built-in protections, namely, the fact that "(Pn.~Qn)" falsifies the generalization. But the problems come when we slip into treating the generalization as a statement about causality.

Consider the two following examples. In the first case, the generalization is one about the sun's position, the moon's position, and an eclipse. The premises leading up to the generalization consist of many years of detailed data about correlations between astronomical positions and eclipses. The generalization is stated in a conditional form, e.g., "If the sun is a position p, and the moon is at p*, then there will be an eclipse tomorrow." In shorthand form, we can say, "whenever p and p*, then eclipse." In this case, it is easy to see how we might get a bit sloppy in our thinking, and move from the 'if, then' conditional to a straightforward causal statement of the sort "The positions p and p* of the sun and the moon are the *causes* of the eclipse." We do this because, *independently* of the correlations which are the evidence for the 'if, then,' we have good reasons to suspect that the causal statement is true. Thus, here, there is no special problem in this example in slipping from conditional statements to causal statements, that is, in blurring the distinction between logic and metaphysics, simply because the causal statement is true, although it is true independent of the evidence for the conditional statement.

However, it is easy to find examples which show the danger of fuzzing over the conditional/causal distinction. Consider the following example. In the 1930s in the United States, a particular phase of drug enforcement took place. This phase has sometimes been called the "marijuana madness." During this period a vigorous propaganda campaign was waged by the government against the use of marijuana. One particular central element of this campaign was the continual use of a so-called "scientific" generalization, namely, "Pot smoking leads to and thus causes heroin use." But this causal statement came about via a drastic misconstrual of the relations between correlations and causal statements, especially in regard to their temporal relations. Phrased more neutrally, it looked as if the evidence pointed to a legitimate generalization that "Heroin use is strongly correlated with pot smoking." But this was interpreted to mean "(x)(Px→Hx)" ("If anyone smokes pot, then that person will use heroin"). Already a miscontrual has occurred, about which more in a moment. However, considered *strictly* logically, there appears to be nothing especially vicious about the generalization. (That is, since it can be falsified by only one instance of a pot smoker who does not use heroin.) Unfortunately, the use of the universalization was not restricted to logical situations alone. Government officials took the misinterpreted universalization and pushed it even further. You will note that there is quite a bit of difference between saying "If pot, then heroin" and "Pot use causes heroin use." The unease you probably feel when you contemplate the difference is appropriate. There *is* a real difference between correlation (that is, between 'if, then' generalization) and causation. Thus not only was the original correlation misinterpreted when it was generalized, but also it was further, and viciously, misinterpreted when it was turned into a causal judgment. 'If, thens' are not causal statements, and ought not to be interpreted that way. But the original misinterpretation occurred when the correlation was first universalized. Correlations, especially when expressed as conjunctions, do not have any necessary logical order which corresponds to the time order in the world. Thus a correlation or conjunction is symmetrical: if 'x' is correlated with 'y,' then 'y' is correlated with 'x.' But this can lead to problems, as is exhibited here. The investigators had researched heroin users, and found that their heroin use was correlated with earlier pot use. But, the time-order differences between "$(x)(Px \rightarrow Hx)$" and "$(x)(Hx \rightarrow Px)$" *must* be reflected in the generalization. The correlation between heroin use and prior pot smoking is different from the correlation between pot smoking and subsequent heroin use.

To complete this discussion, I would like to relate an absolutely knockdown, drag-out example of what could happen were 'if, then' easily allowed to blend into "causes." Although the example is an absurd one, I am sure you will see that it could happen in many similar cases in which its absurdity would not be so evident. The example goes like this. "Every United States victim of lung cancer has previously breathed. Thus, there is an absolute correlation between breathing and lung cancer. Therefore, breathing causes lung cancer."

SUGGESTIONS FOR FURTHER READING

Any introductory book on logic will develop the ideas of this chapter more fully. Copi's book is recommended especially because it is a standard and straightforward account. There are a number of commentaries on the more philosophical aspects of modern logic. I have already mentioned Copi and Gould. One especially good commentary is that by W. C. Salmon, *Logic* (Englewood Cliffs, N.J.: Prentice-Hall, Inc., 1973). A decent collection—but a bit tough at times—is P. F. Strawson (ed.), *Philosophical Logic* (Oxford: Oxford University Press, 1967). A new and relatively idiosyncratic, but worthwhile nonetheless, look at logic is provided by Michael Scriven, *Reasoning* (New York: McGraw-Hill Book Company, 1976). There are many other introductory logic books. None of them is especially dangerous, and so you cannot be really hurt by looking at any of them.

Chapter 3

Theories and Paradigms

DEFINITION OF SCIENCE

Socrates was the first Western thinker to demand that important words be defined prior to their use in discussions. He believed that this practice would solve problems of ambiguity and lack of communication. Since his time, the method of definition has been the approved way to originate discussions. Of course, problems often arise in the very act of defining. It is sometimes quite difficult to obtain agreement about the definitions of words, and this lack of agreement stymies further discussion. In our own century several logicians and philosophers have persuasively argued that it will be impossible to reach single, unique definitions of even the most common terms. The Austro-English philosopher Ludwig Wittgenstein, for instance, concentrated upon the word "game" in an attempt to find some single essential meaning of it.[1] But he found that there simply was not one single concept which captured all the uses of "game." Since

[1]Ludwig Wittgenstein, *Philosophical Investigations*, in Morton White (ed.), *The Age of Analysis* (New York: Mentor Books, 1955), No. 66ff, pp. 230ff.

Wittgenstein's careful analysis, many thinkers have become pessimistic about the chances of reaching agreement about the essential meaning of many of our key concepts, such as "truth," "justice," and "morality." Given this, it is to be expected that definitions of concepts such as "art" and "science" will be equally difficult to reach.

However, even though I must suppose that the pessimists are ultimately correct, and that we will not be able to reach universal agreement about the meaning of the term "science," I am quite sure that universal agreement is not what is necessary in the present context. What we need now is agreement merely about what are some central features of *most* science; such limited agreement will be quite sufficient for our communication. It is quite reasonable to suspect that you might develop doubts as we go along about some of my claims regarding science, but in spite of these I believe we will be able to discuss some of the points effectively.

NATURALISTIC METAPHYSICS

It seems to me that one of the most readily identifiable, indeed, one of the most characteristic marks of science is the fact that its metaphysical system is completely naturalistic. I am sure that many scientists would be upset by my use here of the term "metaphysics." But by "naturalistic metaphysics" I mean nothing different from what you have already seen in Chapter 1. That is, when the attempts at scientific explanation terminate in philosophical statements—those statements which immediately precede the "Just because, that's why" response—the objects and properties ultimately referred to by these philosophical statements are objects which are natural, as opposed to supernatural. Thus, all I am talking about are the ultimate, fundamental beliefs which any general scientific theory must have about answers to the question "What objects are real?" These are the objects which must be hypothesized—not observed—in order for a theory to make sense at any given time of its life. But aside from the term "metaphysics," the really big question here, of course, is: "What do you mean by 'natural'?" An example, plus a definition, should make my meaning clear. First, consider the examples of scientific metaphysical objects we have already seen, namely, the atoms of Newton and the forces of Leibniz. Each of these two very different types of objects is taken to be a member of this world of ours in just the same way that we ourselves are members of the world. In fact, the usual belief is that we ourselves are *constituted*, in some way, by these objects. Since these two types of objects, atoms and forces, are and have been the dominant alternative metaphysical objects since the very beginnings of modern science, I believe that we can safely say that modern science has always opted for a naturalistic metaphysics. But examples are not always completely helpful. Let me attempt a definition.

The simplest way to define *natural* is as I briefly hinted just above: "x is natural if and only if x exists in the same world that we do." But perhaps this more explicit definition is not particularly informative. Let me spell it our more fully. It is clear to us what it means for our own individual selves to exist: We exist just insofar as we interact with other objects like us. This means that we can bump into tables and chairs, hit baseballs, spill drinks down our fronts, and so on. Even more importantly, at least for the present discussion, we can stain blood cells, titrate acids until the indicator changes color, and even turn on electricity in lights and transistor radios. If you compare the sorts of items on the lists in the preceding two sentences, you will see that there is a manifest difference between the sorts of objects mentioned in the two. For the most part, the difference here involves the origination of the concept of the object involved. Objects on the first list—tables and chairs, baseballs, drinks—are common, ordinary, garden-variety objects. And so are the interactions between ourselves and them. But objects mentioned in the second list—blood cells, acids, electricity—are objects which have as their origin scientific conceptualization and discovery. Thus, these objects came into our conscious world in a fashion different from the objects on the first list. Beyond this difference, however, I want to say that there is not much difference between the relations to us of the objects on the two lists. Thus, staining blood cells and titrating acids seem to me not much different from hitting a baseball and spilling drinks down our fronts, at least as far as the interactions between us and the objects are concerned. The interaction is one which, in each case, we ourselves initiate, manipulate, and control (although spilling is rather more often accidental than controlled). Such similarity of interaction seems to me to be especially significant since it argues that we share the same world with cells, acids, and electrical objects, just as we share the same world with baseballs and drinks. Consequently, these objects are natural objects, since they interact with and thus share the same world with us.

It is useful here to contrast these sorts of objects with supernatural objects. Consider gods. Gods always live in a world apart from man—they live somewhere other than our world, whether it be on Olympus or in heaven. Even in those cases—Jesus, for example—when the gods do enter our world, they are not *of* our world. The most important indicator of this distance between our natural world and the world of the gods is the one I just mentioned in connection with the scientific objects: Who (or what) initiates, manipulates, and controls the interaction is what is significant. In contrast to *our* actions in staining blood cells, or turning on the electricity, consider the schemes which mankind has come up with in its attempts to interact with the gods. Ceremonies, religious rituals, and for the most part, prayers play the essential role in man's relations with the gods. Propitiation and supplication in all their guises are the most prevalent modes of attempted interactions initiated by man. Propitiation, of course, is not necessarily *successful*. What this analysis indicates is just the plain simple fact that mutual interactions with these supernatural entities are *not* under our initiation, manipulation, and control. The gods and spirits are under their own control, and not ours. They can just as easily ignore our pleas and prayers. Whether or not we end up interacting with them is at *their* initiation and discretion, not ours. This latter fact necessarily implies that they descend into our world at their will, but we cannot ascend into theirs at our will. Interaction of this sort is not a natural one, in that these beings, or this Being, is not a part of nature in the same way that we are. And this is a great contrast with the cells, atoms, electrons, and what have you of the various scientific metaphysical systems.

Given my account here, it should be clear that the first attempt at a scientific metaphysical system occurred when some parent attempted to explain thunder, lightning, and other common occurrences in everyday, natural terms, and not in terms of entities which were not of our own world. However, to say that some culture's conceptual system contains a naturalistic metaphysics is not sufficient to show that that culture has science, or even is scientific. There are two other central features of science which remain to be discussed.

EMPIRICAL EPISTEMOLOGY

Closely linked to the naturalistic metaphysical status of scientific objects is their empirical epistemological status. Right at the start, however, I should make one thing clear. Empiricist and rationalist scientists were introduced in Chapter 1 in the section on epistemology. But note the difference between *empirical epistemology* and *empiricist epistemology*. An empiri*cist* epistemology is a subclass of empiri*cal* epistemology; thus the classes are not identical. An empiricist epistemology asserts that human perception and sensation is the *sole* origin and source of all knowledge. An empirical epistemology asserts merely that all true knowledge must, at some point, be associated with empirical correspondences and consequences. Thus, some but not all rationalists have empiri*cal* epistemologies, but none of them have empiri*cist* epistemologies, since this latter position denies that some knowledge might originate in reason and thus denies the rationalist position. Just as scientific objects are the sorts of things which can and do interact with us, they are also the sorts of things which we learn about via our sensory and perceptual apparatus. Scientific knowledge, at some point in its construction, must be securely tied into the human sensory system. If it were impossible in principle that some proposed scientific object could ever, under any circumstance, leave a sensible trace in a human sensory system, then that object, no matter what its potential as an explanatory or theoretical entity, could not be considered as a candidate for scientific existence. The necessity for empirical connection is this strong in science.

You might think that my description here eliminates the possibility of there being scientific rationalists. But this is not necessarily so. The point to the empirical epistemology of science is that scientific objects such as the atom and the cell must tie in *somewhere* with human perception. Thus, the rationalist method of making a hypothesis first, and then checking the hypothesis by observation and experiment, is quite consistent with the empirical epistemology of science.

But the epistemological requirement that scientific knowledge be somewhere tied into perception does place some interesting strictures upon scientific work. Consider the following sorts of cases. Many of the very, very small objects which science focuses upon are beyond the reach of unaided human perception. Viruses, for example, or the various kinds of subatomic particles, can never ever be directly perceived by humans. The question thus must be asked, "How can these nonperceptible objects be accepted by the empirical epistemology of science?"

Rom Harré, of Oxford University, has developed a very nice account of the process which admits statements about imperceptible objects through the epistemological gates.[2] Take, for example, he says, the family dog. Now the dog is most assuredly an object about which we can make empirical statements. Thus, "The dog needs a bath" and "Fido has some fleas" are both epistemologically OK. Our unaided, naked-eye observational powers can be used to affirm or deny these statements. Consequently, both are straightforwardly empirical statements. But consider further the flea on the dog's back. If we are careful, and have good light and eyes, we can clearly discern the flea. Moreover, if we look very closely we could even say things like "The flea is blue" and "The flea has a hairy aft end," and verify these statements on the basis of our unaided perception. But let us change the situation very slightly. Just suppose that, instead of using both eyes unaided, we stuck a very powerful magnifying glass between one eye and the flea. In this situation we could close the unaided eye, note the hairy aft end carefully through the glass, and then, closing the aided eye, open the other and verify what we had seen with aided vision.

According to Harré, what we have here is the concept of "extending

[2]Rom Harré, *Theories and Things* (London: Sheed and Ward, 1958).

our natural senses."[3] According to this concept, we are not in any way radically changing the nature of our senses, or the nature of the sorts of information they provide to us. Rather, we are just using some device to amplify or to extend our senses in their normal role. And very significantly, in this magnifying glass example we can use our unaided sense to verify what the aided sense reports. Thus, even though our vision in one eye has the help of the glass in seeing the flea, in the unaided eye it is still completely tied into the normal perceptual situation.

The next stage of Harré's account describes what happens when we interpose an optical microscope—nothing fancy in the way of lenses or lights, just a plain old optical microscope—between the previously unaided eye and the aft end of the flea. In this new situation, using the microscope we might make some particular observation about the hair on the aft end of the flea (something like "It's barbed," perhaps) and then go on to verify this new statement by returning to the glass and using it carefully, with just a bit of strain to the eye. In this manner the new statement becomes as acceptable to the empirical epistemology of science as were the earlier statements verified respectively by unaided and by glass-aided vision. Once again, we have only extended our senses with an instrument which itself can be tightly linked to normal perception. Thus, an intertwining network of sense-extending instruments is built up. The magnifying glass is corroborated by the unaided eye, the optical microscope in turn is corroborated by the magnifying glass, and so on until all the sense-extending instruments are woven together into one firm strand of empirical connection. Insofar as the objects of science can be observed by instruments which fit into the chain linking them to ordinary perception and sensation, these objects are acceptable under the guidelines of the empirical epistemology.

Harré and others, however, have pointed out that not all scientific objects are observable via sense-extending instruments. Thus, even though these objects are completely naturalistic in their metaphysical status, they apparently do not qualify according to the empirical epistemology. Examples of this sort of object include magnetic fields, ultraviolet light, and similar things for which the human perceptual system has no detectors. Sense-extending instruments work on the principle that *aided* human sensation can detect the object under consideration even though *unaided* sensation cannot. Thus, the flea's hairy aft end is not an ordinary, unaided visual object, but nonetheless it is a visual object. Magnetic fields, however, do not interact with any of our sensory modalities. We cannot see, hear, feel, etc., the presence of a magnetic field. No matter how powerful we made our vision, our hearing, our

[3]Rom Harré, *The Philosophies of Science* (London: Oxford University Press, 1972), pp. 21–22.

touch, and so on, we would never be able to detect a magnetic field. Given these facts, it is difficult to understand how magnetic fields, and other entities like them, could possible qualify as empirically observable. I do not wish at this point to go into a detailed discussion of my response to this problem. I will do so later on, when I discuss the "neutrino." But at this point, let it suffice for me to say that it is entirely reasonable to believe that there is a class of instruments which are detectors of things which are not themselves directly observable by human sensation. So long as we can set up a consistent system of rules which regulate and describe these detection instruments, *and which specify the requisite links between detectors and human sensation*, it seems to me that we can allow entities which are detectable to count as empirical. Obviously, I am here presupposing that the entities concerned have already qualified under the requirement that they be naturalistic in their metaphysical status.

At this point, having described the fundamental philosophical issues in the epistemology and metaphysics of science, let us move into a topic which is more specifically scientific than it is philosophical.

The Two Goals of Science

Science, as one distinct sort of human activity, shares many things in common with other kinds of human activity. Thus, there are common features shared between science and history, between science and literature, between science and art, and so on. But in spite of these shared features, what is of interest to us now is to attempt to describe what is *unique* about science, since, to define something (as we are trying to do for the concept "science") is to distinguish it from all other things, and to describe it as it is in and for itself. Thus, what I need to do now is to describe for you an element of scientific activity, namely, its goals, which set it apart from other human activities.

The goals of science are twofold. The first goal is prediction and control. The second is explanation and understanding. Each of these topics will be discussed separately below, but first I would like to say a couple of very general things about the two goals. My first point is that the two goals are not absolutes of all sciences. Some sciences exist today which apparently aim for only one of the two objectives. Behaviorist psychologists, for instance, usually make the claim that they seek only to discover correlations between behaviors, which will be used solely for prediction and control. Most behaviorists then go on to deny that their science will make any attempt to "explain" behavior, at least in the sense that "explain" is usually meant, the sense which involves causal statements.[4] On the other hand, some sciences are almost purely explanatory,

[4]B. F. Skinner, *Science and Human Behavior* (New York: The Free Press, 1966), pp. 29–34. Some of these same themes are also discussed in Skinner's article "Behaviorism at Fifty," *Science*, vol. 140, no. 3570 (May 1963), pp. 951ff.

and do not allow much in the way of prediction and control. The theory of biological evolution is a good example of this kind of scientific effort. Evolutionary theory can describe the course of biological changes in the past, but it cannot be much used to make predictions about the future. As you can see from these two examples, I cannot try to assert in this present section that *all* sciences, equally and universally, exhibit the two goals. Rather, what I will end up describing for you are two objectives or purposes which operate in a very nice overall account of science, but which may be relatively lacking in some particular scientific discipline.

Another general point to be noted is that you might be surprised by the apparent lack of relation between the two goals when we consider certain particular sciences. This is especially true because we bring our ordinary, everyday conceptions along with us as we investigate science. In our typical everyday affairs we usually find understanding and control going hand in hand. Thus, we usually feel that better understanding would lead to better control as a matter of routine. Because of this fairly common situation you might feel a bit of surprise when you encounter the assertions of scientists who seem to believe that the two scientific goals can be divorced from one another. I hope that I will be able to indicate to you why these scientists believe as they do.

My final general point concerns the logic of the two goals. One way to distinguish the prediction-and-control aspect of science from the explanation-and-understanding aspect is this: Prediction and control involve statements which use only correlations, whereas explanation and understanding involve statements which use causal connections. If you will recall our earlier discussion about the differences between these two types of statements, you will remember that some correlations eventually turn out to be translatable into causal connections. A good example of this is the conceptual movement from correlational statements about eclipses to causal statements about eclipses. At first, early astronomers were only able to verify statements about correlations between observed phenomena, for example, the correlation statement "Whenever the sun is at position *p*, and the moon at position *p**, there will be an eclipse tomorrow." But astronomical theory has advanced to the point where this correlation has now become the causal statement "Eclipses occur because of the positions of the sun and moon, among other things." However, as I noted earlier, not all correlations can become causal statements—and therein lies the rub, the distinction between the two goals of science. This will become clearer as we go along in this discussion.

Prediction and Control

In our own times we are completely adjusted to the fact that scientific progress usually means change in our style of life. In fact, it is a commonplace view that the main role of science in society is to make human life richer and easier. But this view of the purpose of scientific activity is a relatively new one in the history of man and science. Francis Bacon first said "Knowledge is power" in 1620. This view sounds completely modern to us, and we have difficulty thinking about a science which is divorced from attempts at practical control. Thus, it is very surprising to learn that it was not until about A.D. 1260 in the works of Peter of Maricourt and Roger Bacon that Western thinkers began to assert the view that there was a possible link between the practical aspects of prediction and control and the intellectual aspects of scientific explanation and understanding.[5] Prior to this era, science, especially in classical Greece and Rome, had been an almost purely intellectual activity, carried out by the intellectual community of the culture. I am not saying that attempts to win practical success over natural phenomena did not occur in these cultures. Artisans and craftsmen certainly were attempting to discover and improve materials and processes which would control nature. But this was not called "science" in those days. The word "science" was reserved solely for the conceptual, or intellectual, activities carried out by the philosophers and other thinkers. What the craftsmen, artisans, and other "tinkerers" were doing went under the name *techne* or, as we would say it nowadays, "techniques" (or even "technology" in a certain sense). Accordingly, these persons were taken to be technicians, and not scientists. The distinction between a "technician" and a "scientist" is an important one, and to a great extent it mirrors important aspects of the distinction between prediction and control, on the one hand, and explanation and understanding, on the other. Let me go into a bit more detail about this.

Control is the most practical concern of all human activity. To control something is to be able to use it for your own purposes. On this account, prediction is the lowest level of control. Prediction about something does not let you control *its* behavior, but at least it allows you to be able to modify your *own* behavior to conform to the predicted behavior of the object. Prediction and control, together, represent the lowest state of scientific knowledge. To see this, consider the following example. In most cases, we can quite securely predict that "For all persons, if x takes aspirin, then x's headache will go away." Moreover, since the prediction is usually true, we can control headaches. But the fact is, we have not the faintest glimmer of an idea about how or why aspirin works. That is, we cannot explain the effect of aspirin, and thus, we do not understand the relation between aspirin and headaches. Consequently, although we know how to predict and control, we do not understand the

[5]Lynn White, *Medieval Technology and Social Change* (Oxford: Oxford University Press, 1964), p. 134.

underlying explanation of why it works. The how and the why knowledge together constitute the highest state of scientific knowledge.

Prediction and control, as I mentioned earlier, involve a logic containing, as does the aspirin statement above, conjunctions—correlations—and their "if, then" generalizations. The discovery of correlations usually is the first step toward creation of a science, and the desire to find the correlations usually arises out of practical necessity. One of the very best examples of this occurred during the earliest days of Egyptian astronomy. Egypt is a desert country for the most part, and yet, oddly enough, it succeeded in becoming one of mankind's first successful agricultural cultures. Egypt's successful agriculture depended entirely on the simple fact that the Nile River flooded every year, and thus released a vast quantity of water and waterborne nutrients into the river plain. But it took an enormous public works project to tame the flood. A mammoth irrigation system was constructed, and then used yearly to capture the floods.

I think that you can already see the problem which produced Egyptian astronomy. The irrigation system was huge, consisting of miles and miles of canals, locks, dams, impoundments, and all the usual irrigation paraphernalia. It takes a vast work crew to make such a system function. Obviously, you cannot keep most of your population standing at the ready, waiting for the flood to occur. But in this the Egyptians were very lucky. The annual flood, although it did not last for a long duration, was extremely regular year in, year out. It occurred very soon—typically within several weeks—after the first day of summer. Thus, if the Egyptians had a good calendar, they could get the crews out to man the floodgates in plenty of time to imprison the flood. But an accurate calendar requires astronomy, since it was only via observation of the movements of the heavenly bodies that early man could accurately tell time. The ultimate conclusion of this long chain of reasons is that the Egyptians were forced to develop astronomy. The history of this development is illustrative.

After centuries of simple observation, the Egyptians had managed to make many correlations between the various positions of the stars and what was happening on the face of the earth. They observed the passing seasons and compared these changes with the changes of the lights in the sky. In particular, they were attracted to the star Sirius. As it turns out, every year just before the onset of the flood, Sirius rises into view at just the instant the sun rises. Thus, the Egyptians were able to formulate the following generalization of the correlation: "If Sirius becomes visible just as the sun rises, then the river will flood, and so it is time to man the floodgates." Since Sirius occupies the dawn position only once each year, this generalization is a very effective predictor.

But it is crucial to note exactly what this conditional statement, this prediction, is *not*. The statement is not an explanation; it does *not* claim any causal relation between the positions of sun and Sirius on the one hand, and the flooding on the other. The Egyptians did not go on to develop any kind of a sophisticated theory to explain why the correlation occurred. Thus, their "science" here was merely of the sort, "At any time, if the stars do *x*, then *y* will occur, and so we should do *z*." In effect, what they were claiming was just, "Whenever the clock of the sky says summer, then the flood is coming; we open up the irrigation system."

The separation between the prediction-and-control goal and the explanation-and-understanding goal which we see here in Egyptian astronomy was fairly typical in ancient astronomy. The Babylonian civilization, which culminated later than the Egyptian, also had a purely correlational type of astronomy. Even the American astronomers, the Aztecs and Mayas, who had developed a calender of astounding accuracy (it was not equaled in the West until the sixteenth century), remained purely correlational. The calendars and other astronomical devices of each of these societies were based upon complex mathematical formulae which, when taken together, allowed mathematical deduction of predictions about future correlated events. These complex mathematical formulae are called *algorithms*, but I often prefer to characterize them as "recipes," since their use and function is identical to those found in a cookbook. That is, most of us—Julia Child being an exception!—do not wholly understand, for instance, why the recipe tells us to use only the yolk of the egg, and not the egg white. We just know that if we follow the complex set of directions—the algorithm or recipe—our actions will be closely correlated with a tasty result.

Many philosophers of science describe the activities related here as "strictly empirical" science. The recipes consist of conjunctions of simple empirical observations, and nothing else. For example, the strictly empirical science of the Egyptians contains the simple empirical observation "The lights in the sky, whatever they are, are doing thus and such (i.e., are rising just to the left of the tower on the horizon, or are setting just behind my cousin's pine tree, and so on)." This observation about the stars is then conjoined to the simple empirical observation "The river is starting to flood" or "The crops are starting to grow." The ultimate empirical generalization of this stage of scientific activity could then be formulated as the statement, "At any time, if the lights in the sky do thus and such, then the river will start to flood." Although this might seem to be an oversimplified case, strictly empirical science of any historical period can almost always be accurately characterized by these cookbook features. Moreover, almost all our present-day sciences developed from this state of knowledge, and have gone on to become the full-blown sciences we know today. Obviously, however, there are sharp differences between a technique such as ancient astronomy and what we now know as "science." The differences between cookbook science and the sciences we know today lie almost entirely in the realm of the function of explanation and understanding, to which we now turn.

Explanation and Understanding

Scientific explanations necessarily involve metaphysical schemes of one sort or another. They must postulate the existence of particular sorts of individual objects, and the interactions which take place between these objects. When the metaphysical scheme of a scientific explanation is wedded to the empirical correlations, this compound produces the kind of total science that we are used to. This sort of transition from cookbook science, from technique, to complete science can be clearly seen by consideration of the astronomy of the Greeks.

The Greeks came later to astronomy than the Egyptians and Babylonians. They were able to make good use of the complex but precise algorithms developed over the centuries of ancient work. The Greeks, however, were not satisfied with mere recipes. They could obviously see that the empirical correlations which had been worked out earlier were highly successful in their job of predicting the course of heavenly events. But the Greeks accepted this success as evidence of an underlying order in the universe. That is, they reasoned that if the lights in the sky, whatever they were, behaved in such a regular fashion that they could be described by mathematical formulae, this fact of their behavior was clear evidence that some sort of real objects were acting according to a knowable pattern. The question raised was thus: "What kind of objects, interacting according to what sort of pattern, would produce the correlations we observe among the lights?" The various answers which the Greeks developed for this question are brilliant examples of human reason in full flight.

The Greeks always opted for naturalistic metaphysical answers. In all cases the lights in the sky were conceived to be natural objects which resembled the earth in many essential features. Thus, for example, Anaxagoras believed that the sun and earth both were natural bodies, one of which, the sun, was on fire. Aristotle, on the other hand, believed that the heavenly bodies were made of different materials than those found on earth. But even given this, he believed that they did resemble the earth in certain important respects, e.g., they had geometrical shapes, and they could (and did) change their physical locations. We find here a full range of theories about the heavenly objects, but even given the diversity of elements in this range, the Greeks' hypotheses (*hypothesis* = a proposal to consider the existence of certain objects, which behave in particular, specified ways) made good use of what they knew about our everyday world. Thus, for example, we might find that the sun, which can be observed *only* as a yellow, hot disc, was hypothesized to be a real, existing, spherical object, similar in some ways to the earth, but on fire. Or another philosopher might hypothesize that the moon was an object similar to the earth. In all cases, the actions of the lights in the sky were the prime data that required postulation of specific actions taken by the objects involved. For example, it is clear that the lights in the sky seem to move relative to our earthly position. The Greeks took this homely fact and from it produced a range of alternative theories which postulated different sorts of motions for the stellar objects. One of these theories has had an illustrious history: Aristotle and several of his contemporaries developed a hypothesis which conceived the earth to be at the center of a series of concentric transparent spheres. The various planets and stars were located upon the surfaces of these spheres. The spheres rotated around their earthly center, but each spun at a different rate. The outermost sphere, for instance, contained all the stars, including those which we know as the constellations of the Zodiac—Aries, Leo, Cancer the Crab, and so on. This sphere rotated every twenty-four hours, and its daily rotation is the explanation of why we see roughly the same pattern of stars overhead at roughly the same time each night (although, of course, the rotation shows a daily change which is small but sufficient to cause the annual pattern that all stars, e.g., Sirius, show up in *exactly* the same place at *exactly* the same time only once each year). The sphere of the sun also rotated daily, which accounts for the observed pattern of night and day. The moon's sphere rotated once each thirty days, more or less. In the final analysis, Aristotle's theory, using a large number of spheres, each rotating in a slightly different fashion, was able to account for all the observed patterns and correlations within a suitable standard of accuracy. Aristotle's theory had many competitors, each of which, through various means, offered explanations of the observations in a manner comparable to his. One theory, that of Aristarchus of Samos, was extremely different from Aristotle's in one respect. Aristarchus' theory proposed that the sun was a major focus of the universe, and postulated that other heavenly bodies rotated around it, rather than around the earth. But Aristarchus, like Aristotle, was able to explain the star movements, night and day, and so on, even though his theory was radically different from Aristotle's.

What we see here in Greek astronomy is a vast range of hypotheses offered by the philosopher/scientists. It might be thought, given this wide range of competing explanations, that the scientific imagination was here operating in a completely free and unfettered mode as it hypothesized. Yet it is clear that there were two strict constraints operating upon the

Greek thinkers in their efforts to explain the observed correlations between the lights in the sky. More importantly, these two constraints operate upon *all* scientific efforts to hypothesize. The constraints are these. First, as I have noted, the objects, behavior, and interactions must be metaphysically naturalistic, and must be such that they provide empirical epistemological significance. This first constraint is simply and straightforwardly the metaphysical and epistemological criterion for scientific explanations. The second constraint is a logical one. The nature and behavior of the objects, by necessity, have to be sufficient to *cause* the observed correlations. Thus, for example, if the observed correlations indicated that the positions of the sun, moon, and earth were related during an eclipse, then the behavior of these objects as postulated by the explanation had to be such that it could be seen to be the cause of the observed eclipse. Beyond these two constraints, however, the scientists were quite free to propose any sort of hypothesis they desired. And as we have seen, this freedom led, on the one hand, to the extreme difference of geometries illustrated by the theories of Aristotle and Aristarchus, and on the other hand, to the similarities illustrated by the proposals of all theories that the sun was a fiery object, that the moon was a satellite of the earth, and so on.

There is a logical point to be learned from our considerations of Greek astronomy, and it is a point which has had tremendous historical significance. The logical point is this: For *every* set of observed correlations, it is *always* possible to conceive a range of metaphysical objects, interactions, and behaviors which will satisfy the two constraints mentioned above. Thus, in any scientific area there is a series of alternative theories which must be decided among. I cannot overemphasize this logical point. *Every* set of observable data has at least two possible explanations.[6] But, obviously, we must believe that not both explanations are true. Consequently, the scientist is every time forced into making a decision between alternative explanations. Decisions, however, are often extremely tricky to make, as I will point out in great detail in later chapters.

This logical point has had a significant historical impact. Due to the difficulty in deciding among alternative explanations, scientists have sometimes made mistakes in their choices among explanations. In this context, "mistake" means that later scientists have come to agree that some other explanation fits better with the observed correlations than the one which they had previously accepted. When this has occurred, a sizable portion of the scientific community during the preceding historical

[6]In strict logical terms, this means that there is more than one valid argument which will imply the data as a conclusion.

era thus has come to be viewed as having been in error, as having confused falsity with truth. The risk of error is what has had historical significance for scientists, and science itself. Many scientists believe that the risk of error in proposing hypotheses is so great that they themselves, and other scientists as well, ought not to propose hypotheses at all. For these scientists, empirical generalizations about observations—what I earlier called cookbook recipes—must be the ultimate attainment of scientific knowledge. To speak in strictly logical terms, they believe that the ultimate statement form is in the 'if, then' universal conditional, and they refuse to move any further toward the causal statement.[7] A clear example of this attitude is to be found in behavioristic psychology, as I mentioned earlier. Behaviorism is the view that scientific statements in psychology ought to refer *only* to correlations between observed behavior states, and do not involve any further statements as explanations for the correlations. A behaviorist statement might thus be as simple as "For any dog, if the dog's pupils contract beyond a certain percentage of their original area, then the dog will either attack or flee." The strict behaviorist will refuse to attempt to link this conditional statement with any underlying explanation, especially one which postulates unobservable metaphysical entities such as "the mental state of anger" or "the mental state of fright." We cannot consequently say, e.g., "The pupil contraction and flight are caused by the dog's mental state of fright." The behaviorist will argue that explanations in terms of causes are unnecessary to our understanding, especially in light of the high risk of error occurring during the process of hypothesizing the existence of explanatory objects and interactions such as, in this case, mental states of fright or anger.

It should be quite clear at this point just how the two goals of science differ from one another. Behaviorists, because of their decision to exclude explanatory hypotheses, limit their scientific goal to simple prediction and control. Other scientists, however, do not hesitate to propose hypotheses to explain observed events. But the risk of mistakes in proposing hypotheses is a real one, and seems to grow in proportion as the number of observation correlations is at a minimum, or at a maximum. From these examples it seems evident that the ultimate kind of science is that one which includes both a high degree of satisfaction of the prediction-and-control goal, and a high degree of satisfaction of the explanation-and-understanding goal. To my mind, it seems quite reasonable to evaluate the various scientific disciplines in terms of these two goals, and to conclude that sciences which highly satisfy both objectives

[7]Some of the essential points in this discussion have occurred to me because I use I. M. Copi's *Introduction to Logic* (New York: Macmillan, Inc., 1972) as a textbook in my logic classes. Chapter 13 is especially relevant.

are more complete than sciences which are deficient in one or the other or both of the goal areas. My evaluative conclusion here is, of course, linked to an analysis of human nature, a human nature which creates science, just as it creates art, history, and all other disciplines. Thus, human life by necessity requires that we be able to predict and control the natural phenomena which constitute ourselves and our environment. But also by necessity, human consciousness demands that reason, that our minds, be satisfied that we understand the objects and interactions which explain the observed natural phenomena. Correlations satisfy the first necessity, and causes satisfy the second.

THE SCIENTIFIC ATTITUDE: HYPOTHESIS VERSUS DOGMA

The final point we will consider in our discussion of the defining characteristics of science involves what might be called the "attitude" of science—which surely raises a psychological point. The psychological point, however, follows closely upon an epistemological point. Moreover, the psychological point does not usually concern particular individual scientists, but rather involves the scientific community at large—which, I suppose, means that the point I have called "psychological" actually turns out to be a *sociological* point. In any case, I will describe the attitude as closely as possible, and give several examples.

First, the epistemological point. Science has an empirical epistemology, that is, an epistemology which ultimately links knowledge up with sensory perception. However, this linkup places a severe restriction upon any particular statement which a scientist claims to be true. Since the time of the French mathematician and philosopher René Descartes, it has been admitted by all philosophers that any given piece of sensory perception, any element of empirical data, is potentially open to mistake.[8] I am sure that everyone is familiar with mistaken perceptions, such as when the creaking board in the nighttime living room is taken to be a burglar, or when the rancher's prize bull is shot as a deer, and so on. Examples such as these may be multiplied to the extent that Descartes' philosophical claim about the potential error-proneness of sensory data cannot be doubted. Because of this philosophical belief, scientists know full well that any particular observation statement, let alone any particular theory statement (given that every set of observations will support more than one theory/explanation), might be false. Thus, the general rule

[8]Rene Descartes, "*Meditations*," in E. S. Haldane and G. R. T. Ross, *The Philosophical Works of Descartes* (Cambridge: Cambridge University Press, 1967), p. 145.

acknowledged by all scientists is "Each and every scientist ought to be ready, at any time, to doubt any given scientific statement."

Now, none of us needs to be a psychologist to realize that this universal rule is a complete idealization. There never has been a human being—scientist or otherwise—who has been willing at all times to doubt any given statement. But it is precisely this attitude which is a logical consequence required by the empirical epistemology. How does real live, scientific practice square with this ideal rule? Actually, it does so in a very straightforward way. Satisfaction of the rule becomes the responsibility of the scientific community at large, and not the responsibility of any particular scientist. You may remember that earlier I described a kind of "trial-by-fire" process which each scientific proposal had to go through before it could become widely accepted. Each hypothesis and observation must in principle be subjected to criticism by individuals other than the scientist who originally proposes the statement. Other laboratories must be able to duplicate the data, make the same observations, draw the same conclusions, and so on. In this fashion the responsibility for caution against error, the responsibility to be doubting of every statement, is satisfied.

Again, of course, the process I have described is an ideal one. However, in the community at large, the actual situation comes quite a bit closer to the ideal than does the individual situation. It is evident that it is easier for one scientist to become convinced that he or she is really observing x, than it is for a whole group of scientists to convince themselves that they are observing x. This is especially true in cases in which x is not there—although, as I gleefully point out later, whole communities of scientists have apparently made mistakes in such cases in the past. However, and this is an important logical point, it is not necessary that the scientific community actually *succeed* in eliminating mistakes in each particular case; what is necessary is only that the community at large have a mechanism which ensures that the doubting attitude is predominant. The public trial by fire is just such a mechanism. The usual name for the attitude which engenders the public trial by fire is the *hypothetical attitude.* The name is used because it reflects the fact that scientific statements are viewed as hypotheses, that is, as proposals of a certain sort, which correspond to data of a certain sort. Since, as we have seen, more than one hypothesis always corresponds to the data, each hypothesis is potentially mistaken. Thus, we believe a statement with a hypothetical attitude when we fully realize that the statement is only one of at least two which might be true. *Provisional* is another term for this attitude. Provisional views are those which are accepted until a better one comes along. But, no matter what you call it, the doubting attitude of science is maintained in order to lessen the chances of a mistaken hypothesis. The problem of maintenance of this attitude over long intervals of time is an interesting one, and one to which we must now turn.

Two Kinds of Traditions

The most basic idea needed to understand the evolution of science is the idea of a conceptual system (which I sometimes also call "conceptual structure"). A *conceptual system* is an entity—a mental object—which contains ideas or concepts as its parts. A concept, to my understanding, is something which can be represented by a written or spoken sentence. Definitions in particular are good examples of representations of concepts. Thus, the definition "The term 'motor vehicles' includes cars, motorcycles, and small trucks" represents some aspects of the concept "motor vehicles." A conceptual system, however, is not just a randomly selected congeries of concepts. It is indeed a *system*, that is, the concepts form a network of logically related ideas. Thus the statement "Semitrucks are not motor vehicles" can be seen to be systematically and logically related to the first statement, since the second is true if the first one is. Two other important aspects of conceptual systems are that the concepts themselves are individually rather precise, and the logical links between them are rather well articulated. To put all the points together, we might extend our two examples to their ultimate limit: The motor vehicle code of any given state forms a conceptual system, that is, an articulated network of logically related, precise ideas.

The logical notions we saw in Chapter 2 can be of help to us here. The concept of a deductive system was defined as a set of given statements (axioms, definitions, theorems, etc.), plus all the statements which could be deduced from the given statements via elementary valid arguments. In the example, we started from various given statements from Einstein's and Newton's respective theories and then deduced various consequences of these givens. If we consider that the fundamental ideas of a conceptual system are analogous, that is, that they play a similar role, to the simple statements and/or defined predicates and names of a deductive system, then the relations between the fundamental ideas of the conceptual system and their compounds and consequences are very similar to the relations which hold between the simple statements of the deductive system and their deduced consequences and conclusions. By this I mean that we can mentally move around in conceptual systems—from one part to another—in a fashion very similar to the ways in which we can mentally move around in deductive systems.

Some philosophers have claimed that conceptual systems in science, theories especially, are acceptable only if they can be arranged in a logical fashion *identical* to a deductive structure.[9] That is, these philosophers require theories to *become* deductive systems in order to be scientifically acceptable. This means that if a theory cannot be set up in a deductive fashion, with axioms, definitions, precisely valid inferences, etc., then that theory does not count as being "scientific." I think that this requirement is a bit too strict. While it is true that some scientific conceptual systems, particularly in physics (kinetics[10] is probably the best example), can be put into deductive patterns, this is not an overwhelming reason to require strict deductive structures for science. Kinetics is a rare case, and it is unfortunate that so many philosophers and scientists have chosen it as the exemplary case of a scientific conceptual system. Not very many other conceptual systems, even in physics, can be put into the rigid deductive mold.

However, after saying all this, I still must claim that the deductive logical system provides a good model, that is, it is usefully relevant, for our understanding of what conceptual systems in science are really like. Thus, although biological or chemical or geological conceptual systems are not arranged in a fashion precisely identical to deductive systems, they are more or less similar. Perhaps it is best to say that a deductive arrangement is a hoped-for *goal* of scientific conceptual systems, but it is not necessary for any particular one to be so arranged in order for us to call it "scientific."

One final point which is important to note is that I have not specified any particular scope to the notion of conceptual system. Thus, although we can call a motor vehicle code of a state a conceptual system, it is also true that the entire set of a state's laws is also a conceptual system, but one which has a wider scope than the motor vehicle code—especially since the code is a small subsystem of the entire set of state laws. Because of this last point, we can say that the state law system contains the motor vehicle code as a subsystem. The notion of scope, of one conceptual system containing another conceptual system, is important for science, as we shall immediately see.

Scientific conceptual systems come in all shapes and sizes. The most familiar and most basic type is the *theory*, a fairly structured entity which includes at least two elements: the set of observation correlations, and the

[9]The very best analysis I have seen of this view of theories is to be found in the "Introduction" to Frederick Suppe, *The Structure of Scientific Theories* (Urbana: University of Illinois Press, 1974), especially p. 50.

[10]Kinetics is the physics of moving bodies. Kinetics can be set up with definitions of two elementary terms ("space" and "time"), definitions of two compound terms ["velocity $\equiv$ space/time"; and "acceleration $\equiv \frac{\text{(space/time)}}{\text{time}}$"], plus three axioms (the "laws of motion," e.g., "$s = 1/2\ at\ 2$"). From this simple set of propositions all the problems of moving bodies can be solved in a straightforward mathematico-deductive fashion.

metaphysical hypothesis which is linked to the correlations. Theories are generally held to be the starting point for more elaborately constructed conceptual systems. These more elaborate conceptual systems may be more or less inclusive than one another. For example, it is believed by many scientists that the conceptual system constituting chemistry is less inclusive than the one constituting physics, i.e., it is believed that physics has wider scope than chemistry. (Another way to describe this is to say that physics is more "fundamental" than chemistry. The notion "fundamental" will be discussed more fully in Chapter 7.) This particular view about the logical relation between sciences has many names, but "reductionism" is the most typical one. The name comes from the fact that it is believed that the statements of chemistry could be *reduced* to the statements of physics, just as certain mathematical statements about fractions can be *reduced* to statements in terms of the least common multiples of the original fractions. (For example, $4/12 + 6/12 = 10/12$ can be reduced to $1/3 + 1/2 = 5/6$.) Many scientists and philosophers, however, believe that reductionism is false. This opposed position does not have any widely accepted name, although I suppose one could call it the "independence movement," since on this view chemistry is logically *independent* of physics. But no matter which of these views is accurate, it should be clear that the conceptual systems of science are of varying degrees of "inclusion," "scope," or perhaps the best term of all, "fundamentality." The point of all this is to introduce you to a kind of conceptual system which is the ultimate within any given science (or, sometimes, group of sciences). This conceptual system is called a "paradigm," and it includes far more than does any particular scientific theory.

The notion of a paradigm was first introduced by the physicist and historian of science Thomas Kuhn.[11] It is a fairly modern introduction which did not receive full currency in the academic community until the late 1960s. Kuhn defines a *paradigm* as a "core cluster of concepts associated with a recognized scientific achievement or set of achievements." It should be clear that paradigms are also associated closely with a particular man or group of men, namely, the one(s) whose achievement it was, although Kuhn does not mention this idea specifically. Thus, one hears talk of Newtonian physics, Darwinian biology, Freudian psycholo-

[11]T. J. Kuhn, *The Structure of Scientific Revolution* (Chicago: University of Chicago Press, 1972). An excellent critique and review of Kuhn's concepts is to be found in Margaret Masterman, "The Nature of a Paradigm," in I. Lakatos and A. Musgrave (eds.), *Criticism and The Growth of Knowledge*, (Cambridge: Cambridge University Press, 1970), p. 59. I must also mention that my notion of a "tradition" is not too far off from Lakatos' idea of a "research programme." See his paper "Falsification and the Methodology of Scientific Research Programmes," in ibid., p. 91. Feyerabend's paper in that same volume is also of some interest for its account of the internal dynamic between the opposing forces of conservation and radicalism in any scientific tradition.

gy, and so on. Kuhn's idea has received some flak, but in its simplest form, it is nonetheless believed by most students of the sciences to identify a significant characteristic of the communal structure of science. Moreover, as I shall now show, the notion of a paradigm is central to explaining the hypothetical attitude of science.

Scientific paradigms tend to have a life of their own, persisting through time, through changing groups of individuals, through advances in measurement devices and mathematical techniques. Paradigms thus tend to become institutional, and scientists who believe them come to practice in what can be called a "tradition." But this is not to say that paradigms never die (nor that they just fade away). On the contrary, paradigms do die, and often, in fact usually, they come to an end in a rapid fashion; at least it is rapid when compared with the length of time that the paradigm existed as the core of a tradition. Kuhn and others use the term "revolution" to identify this period when a paradigm is ending. Let me try to give more familiar nonscientific examples of a "paradigm," a "tradition," and a "revolution."

We are relatively familiar with the philosophical achievements exhibited by the concept structure we call "American government." The *basic* concept structure here, of course, is the Constitution. This document itself consists only of a few short statements which outline a system of beliefs. But as time has gone along, and an ever-changing community of citizens has acted according to the constitutional concept structure, the concept structure itself has grown and changed, accreting bits and pieces of congressional law, administrative practice, and judicial precedent. But each of these accretions is consistent with the basic concept structure itself. Thus, although it is not precisely the *logical* heart of the American tradition, the Constitution remains as the basic element of the paradigm of American government, and it has accumulated a body of logically consistent concepts during the time it has functioned. This whole system, this ongoing intellectual institution, we can safely call the "American tradition" of governmental concepts. So far, I think that the process described here is very analogous to the process of forming a tradition around a paradigm in science. But there is a point beyond which the analogy begins to break down: The Constitution has no concepts written into it to allow its own dismissal or overthrow. It is for this reason that any attempt to overthrow the American tradition of government could be called "unconstitutional." In this particular case, change in the core concepts, namely, removal of the Constitution itself, would end the American tradition. But scientific revolution proceeds differently, indicating a basic difference between scientific and nonscientific traditions.

All scientific paradigms contain concepts which specify the conditions of their paradigm's own demise. When these conditions are met, then the paradigm is rejected, and the tradition associated with it dies out. You might have already guessed that these conditions are none other than those associated with the empirical epistemology. The empirical epistemology is a basic tenet of all scientific paradigms, as I have already pointed out. Thus, each of the theories which are included as part of the paradigm must be tested against, and made consistent with, the empirical epistemology. This is quite similar to the situation in American government. Administrative codes bear the same relation to the overall American governmental concept structure that individual theories bear to their paradigms. Thus, the administrative codes must be tested against, and made consistent with, the Constitution, just as the various theories included within a paradigm must be tested against the empirical epistemology. This has enormous practical consequences. The course of events usually goes like this. When the concepts of the theory—its hypotheses—no longer seem to correspond as well as they should with the growing list of observational correlations, the empirical epistemology demands that hypotheses alternate to the accepted theory must now begin to be very seriously considered, as possible replacements. If the lack of correspondence between concepts and observations becomes great enough, and involves a large number of theories contained in a paradigm, then not only are the included theories affected, but the paradigm itself comes into question. Consequently, in terms of the empirical epistemology, there is an identical condition for rejection of a conceptual system—no matter whether that conceptual system is a restricted one such as a theory, or whether it is a very wide-ranging, all-inclusive one such as a paradigm. Thus, no matter whether it is some small hypothesis in organic chemistry or the gigantic conceptual edifice of chemistry itself, when the observational correlations can be conceived to make better fit with a different conceptual system, then that different conceptual system is the one which is to be preferred.

The question of when to change conceptual systems is an extremely difficult one, and discussion of it is fraught with peril. Indeed, it is one of the most important questions for the theory of science. For this reason, I would like to postpone thorough discussion of the issue until the final chapter. At this point, you need only know that scientific traditions change when the pressure exerted upon a given conceptual system by the requirements of empirical epistemology becomes great enough.

Two Kinds of Revolutionaries: Heretics versus Heroes

The process involved in doing away with adherence to a concept structure, what I call "revolution in the paradigm of a tradition," can be seen to be quite different in science as opposed to other traditions. As our example, let us now consider, not a governmental concept structure but another sort of nonscientific tradition. Consider religion. Religions, like all traditions, are organized around a core cluster of concepts. Judaism, for example, is organized around the Torah; Christianity around the New Testament; Islam around the Koran; and so on. The ongoing communities of persons who believe and practice according to the respective concepts of these documents are each, respectively, members of a particular tradition defined by the concept structure. But religious concept structures can be defined even more closely. Consider, for example, not the all-inclusive Christian tradition, but the more restricted tradition of Roman Catholicism. Roman Catholicism has an extremely well-articulated concept structure, most likely because it has been organized and made precise over nearly twenty centuries. The set of concepts which characterizes Catholicism is called "dogma," and it is this which forms the ultimate paradigm of Catholicism. Dogma, in fact, is such an articulate and precise concept structure that it has been formally codified in a set of books.

An individual is said to be a member of the Catholic tradition only insofar as that person accepts the paradigm represented by the statements of dogma. Someone who publicly rejected some notion which is an element of dogma would be automatically ejected from practice within the tradition. All the organized religions, not just Catholicism, have a name for this sort of individual. Such a person is called a "heretic," and the set of alternate beliefs which he or she proposes as the new concept structure is called a "heresy." If you will allow me to call these individuals "revolutionaries," which I think is certainly justified, we can usefully compare scientific and nonscientific tradition. Contrary to religious revolutionaries, revolutionaries in the scientific tradition are often accorded the status of heroes by the *entire* scientific community at some later date. But revolutionaries in the religious tradition are often viewed as unrecanted heretics by sizable portions of the religious community at some later date, in particular members of the paradigm community they rebelled against. Another way to say this is to say that revolution in the religious community tends to divide, or fractionate, or proliferate the community into smaller subcommunities, which do not agree upon which paradigm(s) to accept. But scientific revolution does not have this effect; in fact, it is quite possible to identify the members of a scientific community as being the same both before and after the revolution.

Luther Let us look at a couple of concrete examples of this. When Luther nailed his thesis (Might we say his "alternative hypothesis"?) to the cathedral door, he was announcing himself as a revolutionary. What he was in effect doing was proposing an alternate set of concepts to those of Catholic dogma. But his proposal was rejected by the Catholic tradition, and since he did not recant, he was branded a "heretic" and kicked out of Catholicism. Luther subsequently set up the initial elements of a new tradition, namely, Lutheranism (to which, of course, he was a "hero"). After Luther, we find that the religious tradition has one more paradigm, and one more community or tradition, than it did prior to Luther. Moreover, it is certainly obvious that a sizable proportion of the wider religious tradition—the traditional Catholics—still consider Luther to be something of a heretic. Analysis of this example allows us to identify two features of nonscientific traditions: (1) rebellion and revolution tend to produce heretics; (2) revolution proliferates competing concept structures.

Einstein Scientific revolutions have just the opposite effect. When Einstein, in his 1905 and 1913 papers, proposed to do away with Newton's theories, theories which had become so basic over 200 years of success that they constituted the sole fundamental paradigm of physics, he was at first greeted with outrage by the traditional community, which is just the way one would expect any psychologically normal group of human beings to react. Such unpleasant greetings are often the case—Pasteur, Darwin, the examples are limitless in number. But the outrage is of a different sort in science. One main difference is that it does not last very long. The force of the hypothetical attitude most often comes almost immediately to the fore in the *community* of scientists, even while *individual* scientists persist in their outrage: This happened at the time of Einstein's challenge. Einstein's theories were rapidly put to the test of conceptual analysis, and were soon enough brought to trial via observation and experiment. It was fairly quickly established that his concepts had at least a surface plausibility. Thus, soon after he had started, Einstein came to be perceived as a revolutionary, but not as a heretic needing to be drummed out of the tradition. Moreover, as time went on and his revolutionary status became more concretized, his proposals were perceived to be not entirely unreasonable. The rest of the story is well known: Einstein's theories came into rapid and wide acceptance, and in the case of the 1913 paper, he came to be perceived as one of the heroes of modern science even before the theories were empirically tested. To describe this in our more formal terms, we can say that the Newtonian paradigm, after 200 years of success, was overthrown by the revolutionary Einsteinian paradigm. In this scientific instance we can see the features of a scientific tradition: (1) revolution does not produce heretics, at least in anything beyond the very short run; (2) revolution does not proliferate competing concept structures. In fact, revolutions in science tend to produce heroes, and unification within the community.

With this analysis, I think that we can conclude our attempts to define science. The points I have made can be summed up in the following way. Science is an activity which takes place in a particular sort of tradition. The tradition is essentially centered in a paradigm, or conceptual structure, which has a naturalistic metaphysics and an empirical epistemology. The goals and objectives of the tradition are twofold, namely, to predict and control phenomena revealed by the metaphysics and epistemology of the paradigm, and to explain and understand these same phenomena. As you can see, this definition encompasses a large number of elements. One of the most significant, however, has not yet had a full enough exposure; I am referring to the paradigm, especially in its role as the focus of an ongoing historical tradition. What I will now try to do is to give you an extended description of the historical development of a paradigm, and, in so doing, prepare you for the material in Chapter 5, which is an in-depth look at the revolution which created modern chemistry.

THEORIES AND PARADIGMS

The Elements and the Ancient Tradition

Modern chemistry was created when Antoine Lavoisier revolted against the accepted paradigm in 1775, a revolution which I will fully describe in Chapter 5. The paradigm against which he rose was called the *phlogiston theory*. It had been successfully used for the greater part of a century, and moreover, combined principles from older traditions, some of which had been around since the time of Aristotle, 2100 years earlier. Given this long vitality, it is clear that we are going to have to go far back in time in order to fully understand what Lavoisier was in revolt against. We cannot fully understand what Lavoisier's revolution was in favor of unless we first fully understand what it was *not* in favor of. This long trip, however, will not be a terribly tedious one, since along the way I can point out some things about our ordinary way of thinking which should both surprise and amuse you. With no further ado, let us begin the trip in the time of Aristotle.

As I mentioned earlier, the Greeks were the first Western thinkers to propose full-blown naturalistic theories as explanations of natural phenomena. Moreover, Aristotle himself had proposed an epistemology which paid special attention to empirical observation—as would any good biologist. What I have not yet described to you is how these features of Greek science fit into their cultural whole. The details are fascinating.

The Greeks from the very first had been intrigued by the problem of understanding and explaining the natural phenomena of change. Since they were an agricultural people, allowed to live mostly outdoors by their temperate climate, they could not avoid being exposed to the patterns of natural change. They were well aware of the changes of the seasons, and of the changes of the days, weeks, and months; and finally, they were especially well aware, as we all are, of the changes represented by biological growth, development, and decline. One of the first of the Greek thinkers, a man named Heraclitus, gave an analysis of the world which still has the ring of truth about it today. In terms of what the senses could observe about the world, Heraclitus claimed, "All is change," by which he meant that no natural phenomenon was stable, permanent, and unchanging.[12] He used the analogy of existence as a river into which one could never step twice. That is not all there is to Heraclitus's views: He also believed in an underlying order—a *logos* or a "logic" of change—but this point was often neglected by his critics. Relative to the empirical world we can still see the sense of Heraclitus's metaphysical claims: Life really does often seem to be nothing but changing, unstable phenomena.

Heraclitus's views about the world of change did not go unchallenged. Another Greek philosopher, a man known as Parmenides, focused upon a different reality as his fundamental element.[13] Parmenides studied human thought and its relation to the world. He noted that our concept of existence, of what it is "to be," was unchanging. This is expressed by language in the statement "Being is"—that is, "Whatever is, is." What we see here is an idea in fundamental opposition to Heraclitus's idea that the main element of the world as we see it is change. On the one hand, Heraclitus perceived the world and saw only change. But on the other hand, Parmenides looked at our conception of reality and saw, in opposition to Heraclitus, that existence is unchanging. The apparent clash between these two views of the world (even though it is based upon a bit of a misinterpretation of Heraclitus) set the problems to be solved by all later philosophers and scientists. From this confrontation between two very different theories about the world developed a complete conceptual system, which formed a paradigm for a tradition that functioned, lived, and grew for almost 2000 years. Let me describe it for you. But first a word of caution.

What I am now going to give you is a composite story of the development of a philosophical view. I have put together parts from a number of different (sometimes, I must admit, even *opposing*) views held by various ancient thinkers and their schools. I am sure many of my colleagues in history and philosophy will object loudly to such a procedure. After all, the views of one man are *his* views, and deserve to be assigned to him personally, and not mixed up with those of his colleagues and critics. This is certainly true. But if I adhered strictly to this precept,

[12]Milton C. Nahm, "Heraclitus," in *Selections from Early Greek Philosophy*, (New York: Appleton-Century-Crofts, 1964), pp. 62ff.

[13]Ibid., "Parmenides," p. 89.

the text you are now reading would be about three times its present length. Moreover, the story I will tell is not entirely a creation from my own imagination—all parts of it did occur. And moreover, regardless of the historical inaccuracy of any of the precise details of my constructions, the composite itself eventually produced a theory not unlike the one I am going to construct. In order to be somewhat scholarly, however, I will identify the origins of thoughts which are significant parts of my reconstruction of the ancient theory. Doing it this way will preserve the essential coherence of the ideas, without entirely diluting them with the necessary tedium of professional scholarship.

The Hidden Structure

The pluralist and atomist schools accepted Heraclitus's analysis of empirical nature as being essentially involved at all times in a process of change.[14] At least, they thought, this was true about our ordinary, everyday observations of the world. On the other hand, these thinkers were quite clear about the validity of the point made by Parmenides. Human thought *did* refer to a stable, unchanging reality. These schools ultimately postulated the existence of an unchanging reality *underlying* the observable world. To use Jacob Bronowski's happy phrase, they conceived the world to have a "hidden structure" of "atoms" or "elements."[15] Plato's analysis focused even closer upon our concepts of the world. He discovered that the idea of a frog, of an oak tree, or even the idea of the most changing of all things, fire, was an essentially permanent mental object. And the representatives of ideas—words—also did not change overnight, or even over a generation. After all, new dictionaries are not needed daily. This idea of unchanging mental objects is an important advance. When we juxtapose the stability of language against the point raised by Heraclitus that only change is real, the tension becomes immediately clear. According to the two themes in this juxtaposition, the world's metaphysical ultimates are: (1) changing physical phenomena and (2) unchanging mental phenomena. Thus, these two totally disparate entities are the only real objects of the universe. But the ultimate impact of this analysis comes only after we are inevitably forced to conclude that there exists a close relation between these contrary objects; namely, the unchanging objects of language apparently *name*, or *refer to*, or *represent*, the always-changing objects of nature. We cannot avoid this conclusion. Language, after all, is *about* the world. But how can this be, how can a set of permanent objects refer to a set of evanescent

[14]A very nice discussion of the pluralists and atomists can be found in W. T. Jones, *The Classical Mind* (New York: Harcourt, Brace & World, Inc., 1969).

[15]J. Bronowski, "The Hidden Structure," in *The Ascent of Man* (Boston: Little, Brown and Company, 1973).

objects? As we reach this point in thought, we realize, as did all these thinkers, that the world and our place in it is far more complex than it may appear. This realization of complexity led to a solution which adopted some of the ideas of all schools, in particular those of the pluralists and atomists.

The logic of the solution is inexorable: If the language and thought are stable, but the ordinary observable world is not stable, then language and thought are not *about* the ordinary observable world. The world must have a *hidden structure*, and it is this underlying hidden structure which guarantees the stability of the set of objects referred to by language. In this move from ordinary observation to postulation of an unobservable but real underlying world, the ancients introduced the Western intellectual community to what I earlier called "reductionism," the philosophical view which hypothesizes that ordinary, everyday objects may be conceptually reduced to their fundamental constituents, and that talk about these higher-level observational objects may be logically translated and reduced to talk about lower-level conceptual objects. An example of reductionism is the common assertion that "Human beings are *in reality* only worth 98 cents, since that is the value of the chemicals which constitute us; after all, we are, *in reality*, metaphysically speaking, only a batch of various chemicals, since all our organs, tissues, cells, and so on can be reduced to their constituent chemicals." Although this assertion needs to be corrected for inflation, it is still clear that it exhibits a particular view about what human beings *really* are, namely, chemicals, as opposed to what they *appear* to be, namely, organs, tissues, etc.

Different schools held varying views about the details of the reduction of ordinary objects to their fundamental objects. In general, however, the logic, or style, of all the moves is typical. We start from the idea that our ordinary observational and perceptual processes are central features in any solution to the problem of finding the hidden structure which is the ultimate referent of language. What we then attempt to do is to find the underlying fundamental features of perception, and to see how these correspond to everyday observation. Thus, the perceptual system is focused upon, in an effort to isolate the basic functions and data which are present in human sensation. From this analysis is developed a systematic account of the most fundamental data which humans could get from perception. The final belief was that the ultimate level of the observational world is the "hot," "cold," "wet," and "dry." This belief may sound strange, but it is not really strange. If you ask "What can we observe?", one usual answer is chairs, table, houses, trees, and so on. But this answer must be rejected, since these observables are always changing even while our concepts of them remain the same. There is, however, another plausible answer, although it somewhat distorts the normal sense of the term "observe." If we ask "What can we observe?" in a context which involves noting fundamental data produced by the senses, then we can say that we observe sights, sounds, colors, and so on, with our sensory organs. This is the kind of answer that many ancients gave to the question "What can we observe?" In the fundamental sense, according to this theory, what we can observe is things presented to us as clusters of complexes of sensations of hot, cold, wet, and dry. All other sensations and ideas are derivative upon these four basic qualities. "Snow," for example, might be compounded out of many sensations, primary among which would be sensations of wet and cold.

But the ancient philosophers did not leave scientists locked up in their own minds, observing sensations as though they were a TV program of hots, colds, wets, and drys. Rather, these philosophers believed that certain real, ultimate physical features of the natural world corresponded strictly to these observable qualities. In the details of this view, which was propounded first by the pluralist philosopher Empedocles, there were four types of basic physical structures, each linked to two of the basic qualities. Thus, hot and dry were the observable counterparts of the basic structure "fire." (See Fig. 3-1.) In one sense, the qualities "hot" and "dry" *constituted* fire. But in another sense, more significant for our account, "hot" and "dry" were the perceptible parts of the fundamental structure "fire." "Cold" and "wet" were the perceptible aspects of "water." And so it went for all four. These four basic structures—air, earth, fire, and water—were Empedocles' elements, as they later were to serve Aristotle as well. It was from these basic elements that all other objects, such as rocks, frogs, and even people, were formed. In this way the contradiction between the stability of language and the instability of the world was

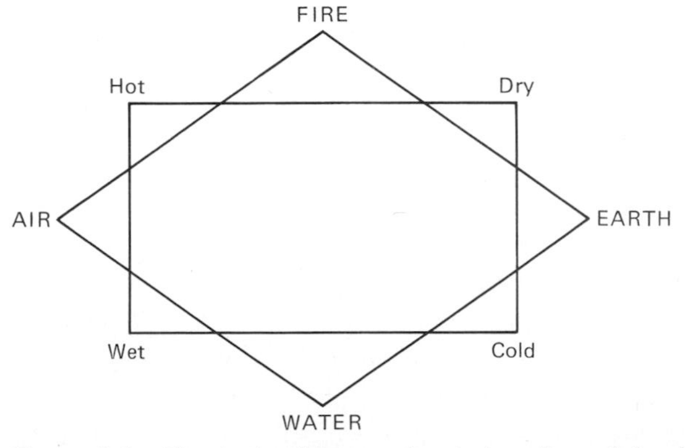

Figure 3-1. Physical entities are located on the points of the rhombus, and sensible qualities are located on the points of the rectangle. Each physical form corresponds to the concatenation of two sensible qualities. Fire, for example, corresponds to the qualities hot and dry.

relieved: The usual observable world of growth, seasons, etc., was unstable, but underneath, in the hidden structure, was the permanent, unchanging world of the elementary forms air, earth, fire, and water.

A word about terminology here. In what follows, I will persist in calling air, earth, fire, and water "forms." By "forms" I simply mean the ultimate patterns, or orders, or structures, which exist in the physical world. This usage is somewhat controversial. Plato had been the first to use the term "form" in a metaphysically significant fashion. He used it to refer to those objects which provide the patterns for the things of the natural world. But from this point on, we must turn our eyes to Aristotle, who used the term somewhat differently. The "form" of something was also its "pattern," which can range in meaning all the way from the simple shape of a rock to the essence of a living being (not unlike the way in which the genetic code of a human being is in some strong sense its "form"). In the most restricted interpretation of ancient thought, the hot, cold, wet, and dry were the ultimate forms in which objects could appear. But certainly, at least in a physically meaningful sense, the elements—the fire, earth, air, and water—of which a thing are composed are its forms, that is, the things from which it can be conceived to be formed. It is this primitive sense of "form" which I think is relevant here. Thus, while Aristotle did not himself provide the analysis I will now give, I believe it is fair to push his concepts to the point where they include just the sort of elementary description I will give you. The importance of all this from the point of view of "paradigm" and "tradition" is easy to see: We are still under the influence of Aristotle's view today. The concept of the "form" of something has passed from the technical philosophy of Aristotle into the ordinary thought processes of us all. Let me give a couple of examples of this.

Formulae

Everyone is familiar with the concept of "formula." A *formula* is a sort of road map, or a detailed description of how something is organized or produced. Formulae, especially in science, are taken to reveal the most intimate hidden structures of things. Hence, we have chemical formulae of water, gasoline, and most importantly, Coca Cola. Think how intimate and significant the formula of Coca Cola is: It reveals the secrets of the substance. This shows us the reason why the formula for Coca Cola is so closely guarded. If you ask yourself, "Where did the concept of a 'formula' come from?", the answer should already be obvious. "Formula" is the symbolic representation or description of the forms which make up a substance. Thus, if a substance is constituted by air, earth, fire, and water, then the formula of that substance will be the list which describes the various ratios in which air, earth, fire, and water are combined to make up that substance. Hence, even in the very concept of a formula, we are presupposing the ancient theory which asserted that every object consisted of ratios of the elementary forms. Note one other important thing, however. Since there was a correspondence between the sensible perceptual qualities hot, cold, wet, and dry on the one hand, and the physical elements air, earth, fire, and water on the other, it follows that the scientist could give a formula in terms of either set of entities, either physical or sensible. This possibility will be of importance to our understanding of how the ancient—especially in its Aristotelian aspects—theory was later developed into the paradigm used in alchemy. But let me first cite another example of how the older theory passed into ordinary conceptions.

Transformation

Attendant upon the notion of a formula is the concept of changing the formula; that is, of altering the pattern of the basic forms of a substance. The name of this process could not be other than *transformation*. Again, we must remark the presupposition of the ancient theory in this concept. Thus, if substances have forms, and hence, formulae, then chemical change consists in the alteration of these basic forms—a process which must be called transformation.

Thus, the tasks of the Aristotelian scientist were clear: to discover and understand thc basic formulae of the natural substances of this world, and to be able to list their basic forms and the formal patterns of their hidden structure. With this magnificent conceptual tool at hand, Aristotle and his successors went on their way to develop just this task. I must point out that, for these scientists, practical control was not the goal; rather, their goal was to apply the conceptual system of the formal philosophy in a massive attempt to understand the world. Thus, although the attempt to explain the world in terms of the hypothesis of the four elementary forms is a vast undertaking, an undertaking which we applaud even while recognizing its impossible scope, we must at the same time realize that this undertaking cannot straightforwardly be called science in terms of our earlier definition. The formal philosophy, vast as were its goals, did not orient itself toward practical control. Thus, even though the first Aristotelians believed an empirical epistemology, they did not link this epistemology to practical control as a goal. Consequently, they missed opportunities to correct their metaphysical hypotheses as provided by the experimentation necessitated by practical control. This deficiency, however, started to become rectified by latter-day Aristotelians, philosopher/scientists of the thirteenth and later centuries. It was these, the alchemists, who attempted to apply the later Aristotelian system in the actual physical world.

The Alchemists

In the above, I can somewhat justifiably be accused of pushing Aristotle and the other ancients too far, of overinterpreting their words to present a clear and coherent picture. However, in respect to Aristotle's view in particular, it is clear that even if he himself did not say quite so much as I have claimed, then certainly his successors did. In fact, one particular group of his successors is notorious in the history of science. This group is called the alchemists, and they are usually portrayed as charlatans, magicians, and wizards interested only in their own financial gain. Certainly there were humbugs, quacks, and medicine-show performers among the alchemical ranks. But a bad press should not obscure the truth of the matter: Among the alchemists were men of goodwill, intelligence, and real scientific ability. These men believed honestly in the theory of alchemy which had been developed from ancient hypotheses, and they carried out their experiments under the guidance of the theory, just as other scientists must always be guided by their own theories. The question we must ask ourselves, then, is *not* "Why were the alchemists such quacks?" but rather "What are the significant theoretical elements which guided the alchemists?"

The alchemists are most famous, or infamous, for their attempts to turn lead into gold, which sounds crazy to us now. But the reason that it sounds crazy to us is that we happen to believe a different theory of hidden structure—a theory which, we hope, is more truthful as well as more fruitful than the theory believed by the alchemists. However, given their theory, transformation of lead into gold was quite a reasonable thing to expect to be able to do since the theory implied two principles to follow in experimentation, and these two principles involved lead. Let us see why.

As I pointed out above, in the older concept structure, substances can be compared according to two methods. In the first method, we can simply list the basic physical formula, giving the ratios of the combination of air, earth, fire, and water which are present in the substance. However, because these substances are fundamental and, in a sense, hidden, it is not always possible to accomplish this. Thus, we must use the second method in these cases: comparison of the sensible qualities of the substances under consideration. This method works only because there is a close and dependable correspondence between the physical form and the pair of sensible qualities it affects in us. Thus, if something is hot and dry, then it must contain fire, and so on. But the ancient reductionist hypothesis goes even farther than this. Since higher-level structures are compounded from more basic ones, e.g., cells are compounded from chemical structures, it follows, in an epistemological sense, that certain observable qualities are also compounded from basic ones. For example, the "shininess" of metals, which is a complex observable property, corresponded to a complex physical compound containing at least fire and earth. Using this method, it is possible, according to the theory, to deduce the atomic constitution of substances we cannot examine at the atomic level. It was this approach which was used by the alchemists, at least at the beginning of their enterprise. Their approach can be summarized in the principle that "observational analysis reveals physical formulae."

A second principle is linked to the one above: If a chemist succeeds in altering the physical formula of a substance, the transformation must be revealed by observable changes in the qualitative features of the resulting product of the transformation. Thus, if fire is added to an earth (ore), a shiny metal results. The two principles can be put together to state: There can be no physical changes without qualitative observational changes and vice versa. It was this rule which got the alchemists into their unproductive backwash of scientific progress.

Why did the alchemists think they could transform lead into gold, and most importantly, why pick on lead and not silver or something else? Basically, their belief is linked to the two principles expressed above. Let us construct a hypothetical comparison of lead and gold, focusing upon the observable qualitative analysis outlined in Fig. 3-2.

As you can see, lead and gold are qualitatively very similar. But in accordance with the first principle, this necessarily implies that they are physically—atomically—very similar as well, since observable qualities correspond to physical structure. Hence, lead is the obvious choice simply because it is so very similar to gold. That is, the transformation of a substance very much like another is obviously going to be much easier than if one starts from substances which are very dissimilar. Imagine trying to convert water into gold, or wood into gold. An inconceivable task! But the theory goes even further. The qualitative comparison between lead and gold makes evident a very significant guiding point: The main differences between gold and lead concern shininess and color. Lead is dull and gray, gold shiny and yellow. Now, since what one wants to transform is just these observable qualities, namely, to change dull to shiny and gray to yellow, then it is clear that those elements which will

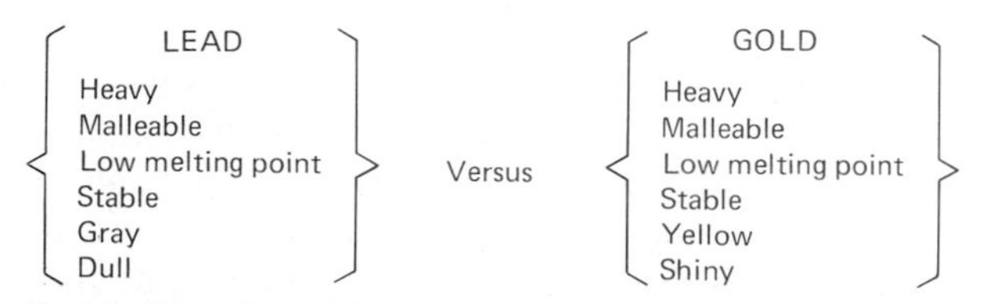

Figure 3-2. Comparison of the gross observable qualities of lead and gold. Note that the two metals are very similar in gross appearance, which implies that they are very similar in fundamental physical constitution.

produce these qualities must be added to lead. Obviously, fire is just what we are looking for, with its yellow color and fiery gleam. Thus, the predicted procedure according to alchemical theory is to add or at least involve fire in the transformation of the dull grayness into shiny yellowness. The theory is not quite good enough to tell you *exactly* what and how to do, but at least it tells you to use fire in the transformation. Consequently any process which the alchemist knows will add fire to the lead should be experimented with.

It is clear from this analysis that alchemy is entitled to be called a *scientific* theory. This does not mean that alchemy ever achieved its goal as a practical science, since the alchemists did not succeed in transmuting lead into gold (modern physics can do the job, but the cost is prohibitive). Since science has the twofold goals of practical control and intellectual understanding, we can safely conclude that alchemy is a science which is immature because it achieved the latter goal without much success at the former one. On this account, alchemical transmutation attempts can be considered to be unsuccessful experiments directed toward achieving practical control as guided by the metaphysical hypothesis of the four forms, and by the empirical epistemology which linked the four forms to their observational qualities—hot, cold, wet, and dry. Clearly, then, alchemistry is not magic, but misguided and unfruitful scientific theorizing. The effects of the ancient theory, however, did not stop with the transmutation experiments in alchemy. Aristotle's theory was developed to the point where it could be deployed in medicine and psychology, as well as in other areas of physical science. Thus, the ancient theory passed from being a conceptual system with broad scope, into being of such wide scope that it served as the main paradigm for almost all Western intellectual traditions, including the religious, political, and aesthetic traditions in addition to the scientific. A couple of more examples from the scientific tradition will give good evidence of the significance of a paradigm for a tradition.

From Theory to Paradigm

Deployment of the ancient theory into medicine resulted in the development of concepts which we find operating in our ordinary patterns of speech and thought even today. While the four forms, air, earth, fire, and water, were the basic elements in the development of alchemical theory, it was not until later that they became significant factors in the development of the theoretical biochemistry of the human body. During the time of Hippocrates, Aristotle, and other classical Greek thinkers, certain medical theories had become linked to the hot, cold, wet, and dry metaphysics. These theories involved four humors, or bodily fluids. Galen, the classical Roman physician, had further refined these concepts, and proposed that they were related to Empedocles' four forms. What we see during the later medieval period is the linkup of two very different scientific theories, medicine and alchemy, via their identical metaphysical schemes. Both scientific theories relied upon the metaphysics of the four forms, and consequently, correlations could be found between alchemical elements and medical humors through the mediation of the four-forms metaphysics. In medicine, each of the four alchemical elements became associated with one of the four bodily humors which were conceived to be the media responsible for all internal processes. Air, for example, was taken to be strongly influential in the processes having to do with blood. Fire, on the other hand, was responsible for the activities of the choler, or yellow bile. Earth and water were similarly associated with their respective fluids. (See Fig. 3-3.) The relationship between overall general bodily behavior and the four fluids was correlated in interesting ways with the ideas we just saw in alchemical theory. Each person, for instance, was understood to be medically describable in terms of his or her present balance of the humors. Moreover, as implied by correlation to the two epistemological principles of the alchemical theory, the humoric balance was describable in, and reducible to, terms of the ratios of air, earth, fire, and water, on the one hand, or hot, cold, wet, and dry, on the other hand.

Health and disease are quite easy to define in terms of this theory. Health is simply that state of being in which the four humors are stably balanced. Disease is a permanent or temporary, and sometimes acute,

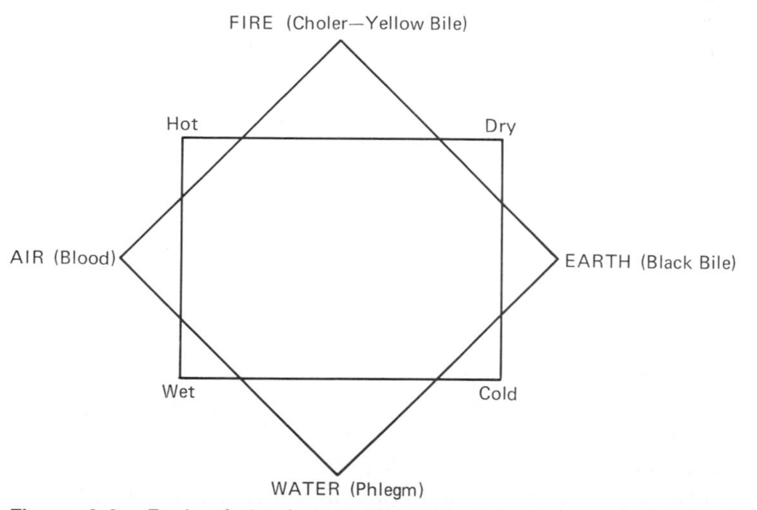

Figure 3-3. Each of the four bodily humors became associated with one of the alchemical elements. The logic of this association involves the fact that both humor theory and alchemical theory are based upon the four-form metaphysics of hot, cold, wet, and dry.

imbalance among the four. Fever provides a good example of how this analysis works. When people have a fever, it is quite evident that they are hot to the touch, and in fact, they are often said to be "burning up." Even cursory reflection upon the observable signs of fever indicates that the person involved has too much of the choleric fluid and, ultimately, too much fire present in the bodily operations. One useful point of the theory is that it clearly allows deduction of treatments required for the various diseases. Fever, acute excess of choler and fire, requires that its opposite form be administered. Thus, water may be used in many ways to "put out the fire." Methods include forcing fluids, especially those such as fruit juices which are high in water content. Additionally, since the victim is too hot, cool water baths and so on will be useful.

In modern medicine, some diseases still retain their intellectual connections to this ancient theory. Cholera, for example, is the modern name of a disease which is characterized by a high fever. It derives its name from the ancient Greek word *choler*, for bile. In another case, we all know that colds and viruses are often accompanied by excesses of phlegm, which is another of the four humors. Obviously, one who has a cold and its attendant runny nose would be far better off if remedies were administered to dry up the excess phlegm. Accordingly, hot things could be administered—pepper concoctions and so on. Health would return with the reduction of the excess humor. But the most interesting thing about colds involves our modern concepts about their origins. According to the ancient theory, phlegm is associated with the two perceptual qualities "cold" and "wet." Logically, then, one main way to catch a cold is to get too cold and wet. But this deduction does not hold true solely in ancient theory. Even today we tend to associate colds with being cold and damp, an association which is a complete throwback to ancient theory, in light of the fact that colds are caused by viruses and not by being cold and damp. It seems pretty obvious that the name "cold," as used for the virus disease most of us know all too well, is ultimately linked, not to modern medicine, but to a medicine of the far past which links the sensible qualities cold and damp to the disease, via the notion of phlegmatic excess.

These are only some of the many examples of how the ancient theory still pervades our modern, informal ways of thinking and speaking, in disciplines which have disavowed the theory from a strictly formal and scientific point of view. Thus we see that, although medicine no longer believes the older theory—it has not since the nineteenth century—the theory devised by Aristotle, his contemporaries, and their successors had such vast success in explanation and understanding that it has percolated throughout our fundamental concepts of disease and health. The effects of these percolations continue long past the demise of the scientific use of the theory. In this example from medicine we see how a concept structure—the basic physical theory devised by Empedocles, Aristotle, Hippocrates, and the other ancients—gained far wider scope than it initially had in alchemy. But alchemy and medicine are not the final limits of the older concepts. It was in psychology that these concepts reached their ultimate extent.

Diseases may be defined in the ancient medical/physiological terms as acute imbalances in the four humors. But bodily symptoms are not the only manifestations of such imbalances. The humoric imbalances also have psychological aspects. Thus, when one has a fever, one reacts psychologically by being irascible, bilious—in a word, choleric. When one has a cold, one is turgid, languid, slow-moving—in a word, phlegmatic. And so it goes. But the psychological manifestations of the humors do not occur only during diseases. The older theory in addition has a strictly psychological/behavioral interpretation of them.

Medical consequences come from *acute*, or short-term, gross imbalances in the humors. Psychological consequences, on the other hand, are produced by *chronic*, long-term, minor imbalances among the humors. On this analysis, individual personality types are closely correlated with the dominant chronic imbalance. Four basic types are distinguished: the choleric, phlegmatic, melancholic, and sanguine. These types are associated, respectively, with the predominance of choler, phlegm, black bile, and blood. Choleric people in general are irritable, quick tempered, hot-blooded, just as we would expect people to be with too much fire in their bodies. In modern parlance, we still say that the choleric person is "fiery." On the other hand, phlegmatic people are, as we earlier saw, slow-moving and rather tractable. Juxtaposed to the phlegmatic person is the melancholic. *Melancholic* comes from the Greek words *melas* meaning black and *choler* meaning bile, and it is an excess of this humor that generates the gloom and depression of melancholy people. The fourth personality type is the sanguine. Sanguine people usually have cheerful, ruddy faces, and it is the lively coursing of blood through their veins which is responsible for their abundant vitality and energy. When such energy is lacking, we of course say that the person has "tired blood." In an almost incredible fashion, these ways of speaking have lingered on from ancient medicine, and still resist elimination in favor of more modern concepts. But these holdouts do not exhaust the list; there are many others. Let us consider just a couple more.

We often speak of certain national and racial groups as having specific personalities. The Latins, for example, are often characterized as fiery or hot-blooded. The Scandinavians, on the other hand, are often taken to be phlegmatic and stolid, not easily aroused to emotion. It is easy to see how our ancestors justified their belief in such stereotypes on the basis of their scientific explanations. The environment of the Latin nations—the Mediterranean basin—is extremely rich in fire particles coursing out from the sun. The constant presence of high-intensity sunlight can be reasonably expected to modify the chronic balance among the humors of all Latin individuals. Thus, no matter what the original humor ratio of any particular individual might have been, in the long run this ratio must be overwhelmed by the constant, intense flux of fire particles streaming from the sun.

The phlegmatic Scandinavian personality, on the other hand, is easily explained on the basis of the consistently higher proportion of cold, damp, rainy weather which plagues the northern lands. Additionally, of course, the lack of intense, hot sunlight contributed to the chronic excess of phlegm in the Scandinavian individual. There can be no doubt about the power of the ancient paradigm to explain the purported facts represented by these stereotypes.

The final example I wish to point out for you is one which I am sure everyone is familiar with. Think how often we say, of someone who has gotten out on the wrong side of the bed, "He's in a foul humor today!" or of someone who is in fine fettle, "What good humor she's in this morning!" These expressions are absolutely basic to our descriptions of our human moods. But the explanatory power of these descriptions—in other words, their capacity to inform us about the underlying basis for the moods—is almost nil because it is lodged in a scientific paradigm which no longer is accepted. Thus, although the older scientific paradigm has been superseded in modern sophisticated thought, its conceptions still remain at the bottom level of many of our ordinary, everyday thoughts and descriptions of the world.

Let me now briefly summarize what we have learned in this chapter about theories, paradigms, and traditions. Theories are conceptual structures which are produced in order to fit a fairly well-defined, limited context. Thus, the air, earth, fire, and water theory was developed for and applied within a fairly narrow range of physical and chemical phenomena. However, as we saw, the theory did not stay within its restricted area. Rather, explanations for widely divergent areas such as medicine and psychology were developed and came into widespread use. This use over a long period of time, over a wide scope of phenomena, defined a tradition, namely, those communities of practicing scientists who used the ancient concept structure in explanations of work in their disciplines. Obviously, it is impossible to know exactly when a theory has turned into a paradigm. We cannot be expected to agree unequivocally about what width of scope is necessary for a theory to qualify as a paradigm. But universal agreement is not needed for our purposes, since we need only to be able to understand some clear examples of paradigms as they are presented. In any case, it is plain that the ancient theory grew in scope as time went on, and equally plain that this ever-growing concept structure defined a community of scientists whose scientific efforts were organized around the explanations provided by the theory. Finally, it is clear that the paradigm was of such great scope, and had such impact upon men's minds, that some of its concepts persist even today, long after the paradigm itself has been overthrown.

At this point, the question for the next chapter should be evident to us all: If paradigms have such impact upon—if they hold such sway over—the minds of scientists, then how can any paradigm ever be replaced? In other words, how can there be any scientific revolutions? The answer to this important question revolves around the notion of a scientific "discovery," and how discoveries affect science.

SUGGESTIONS FOR FURTHER READING

An excellent discussion, replete with historical analysis, of the two functions of science (explanation and control) can be found in the first chapter, "Law and Cause," of Emile Meyerson's *Identity and Reality* (New York: Dover Publications, Inc., 1962). Still the most readable account of Greek science is Benjamin Farrington's *Greek Science* (London: Penguin Books, 1953). Farrington pushes his own theory heavily in the book, and although more recent scholarship has cast some doubt on the viability of the theory, the book is still quite readable and understandable. The most thorough book on the subject of ancient astronomy continues to be J. L. E. Dreyer, *A History of Astronomy from Thales to Kepler* (New York: Dover Publications, Inc., 1953). Sir Karl Popper has been the foremost proponent of the antidogmatic necessity for science. See any of his works; one of the best technical presentations is to be found in his *Conjectures and Refutations* (New York: Harper Torchbooks, 1965). On the subject of alchemy, a long-time standard has been John Read, *Through Alchemy to Chemistry* (London: G. Bell and Sons, Ltd., 1961). Finally, I should mention that the idea of the network of relations in a concept structure has been developed with a great degree of elegance in W. V. O. Quine's essay "Two Dogmas of Empiricism," in his book *From a Logical Point of View* (New York: Harper Torchbooks, 1961).

Part Two

Scientific Discovery

Science is famous (in some cases, such as the atom bomb, "notorious" may be a better word) for its discoveries. Can anything philosophically relevant be said about discovery in science? Many philosophers have thought not. They have believed that scientific discovery is like artistic discovery or mathematical creativity, etc. According to these persons, discovery/creativity is a subject best left to examination by the psychologist or psychiatrist. Obviously I disagree; otherwise this second part of the book would not exist. My belief is that scientific discovery can at least be illuminated as to the intellectual elements which led up to the discovery in question. I will start out by discussing a particular riddle proposed long ago by the philosopher Plato. This riddle ends in the apparent conclusion "Discoveries are impossible." By laying out the logic of Plato's riddle in Chapter 4, I will set up the framework for analysis of the discovery question.

But discovery is a highly *personal* thing: discoveries are made by individual scientists, in individual and unique historical situations. Thus, any thorough discussion of discovery must include reference to these personal elements. In Chapters 5, 6, and 7, you will be introduced to three scientists. Incidentally, each of these scientists is a bit of a hero to me, and I will tell you some of the personal reasons for my hero worship. Lavoisier's courage in opposing a very strongly entrenched system of concepts, Pasteur's exquisite use of a weak data base collected from his activities in wine and beer making, Pauli's attendance at a dance instead of personally reading his proposal which subsequently turned the scientific world on its ear—I will show you these aspects of the human side of science and share with you my sense of delight in them.

But obviously not all, or even most, of the detail of Chapters 5, 6, and 7 will be personal biography. Rather, large amounts of discussion will focus on scientific data precisely because, if you do not understand the scientific detail, you will not be able to understand what the discovery is all about. In the discussion of Lavoisier (Chapter 5), you will come to understand how oxygen was "invented" and the previous substance, phlogiston, ceased to "exist." I realize that it must seem peculiar to you to think about oxygen being "invented," but that is one main way to think about many scientific discoveries. That is, these discoveries are not like Columbus stumbling across a new land; rather they are more like discovering how baseball can be invented out of a previous game such as British rounders.

Chapter 6 will take a closer look at Pasteur's discovery ("invention"?) of the role played by bacteria and other microorganisms. I will make the claim that Pasteur should be much more respected for this theoretical discovery, the discovery of the role of microorganisms, than for his more practical discoveries such as vaccination for rabies or pasteurization of milk, beer, and wine. At the end of the chapter I will take a quick look at the connections between biology and chemistry/physics. Some philosophers and biologists believe that some day in the future, biology will not exist as a science separate from chemistry/physics. I argue against this proposal.

Finally, in the last chapter of Part Two, I will get into a discussion of Pauli and his discovery—the "neutrino," the subatomic particle whose name means "little bitty neutral one" in Italian. Pauli must definitely be said to have invented the neutrino. And the announcement of his invention produced a vigorous opposition, precisely because this little bitty neutral particle would and could never be directly observed. Thus, it looked quite a bit as if the neutrino were only a figment of the young Austrian scientist's imagination. But I will show how and why Pauli *had* to make his invention. It will be clear that had he not made his proposal, then an extremely and crucially fundamental part of modern physics would have had to be thrown out. After introducing this argument, it will be necessary for me to get into a fairly long discussion of what "fundamental" means. In this discussion, some significant features of the logic of scientific thought structures will become evident, and so the length of the discussion will turn out to be worth the effort.

After all this, my analysis of scientific discovery will be finished. You will have met some fascinating human beings, and seen some incredibly brilliant reasoning. And you should have a better idea about what goes on in scientific discovery.

Chapter 4

Plato's Dilemma

INTRODUCTION

Many recent theories of science have attempted to divide scientific activity into several phases, chief among which are the discovery phase and the verification or acceptance phase. During the discovery phase, according to this view, new data and/or hypotheses are introduced to the scientific community in an effort to convince the scientists that the new ideas offer significant improvement over older concepts. The verification or acceptance phase is just what its name implies: It is the period during which the new ideas are carefully tested, scrutinized, and put through the trial by fire.

While this view of scientific activity has provided some very useful insights, especially with regard to the logical methods which function during the two phases, it has not proved to be especially enlightening with regard to analysis of the historical and psychological currents which pervade science as a human institution living through time. A better viewpoint seems to me to be provided by the notions of paradigm, tradition, and revolution which I introduced in Chapter 3. Given the complementary benefits of the two different analyses, my goal in this chapter and the next three chapters is to attempt to put together these two different analyses of scientific activity; that is, I will attempt to combine the discovery/verification theory with the theory involving paradigms, traditions, and revolutions. The procedure I will follow is a fairly straightforward one. First, I will say a few very general things about discoveries in science, and set up a logical puzzle involving the problem of how discoveries relate to the historical aspects of ongoing traditions. Then I will launch into the main element of Part Two, a presentation of my own view about what went on during three very important scientific revolutions.

In Chapter 5 I will describe the eighteenth-century collision between the revolutionary new chemical ideas of Antoine Lavoiser and the traditional views held by Joseph Priestley. In this fierce interchange between a young Frenchman and an older Englishman we will find the origins of modern chemistry, which can and must be dated from the events we shall study. Following this we will look closely at another very important revolution, this one in biology. In the nineteenth century, Louis Pasteur's revolutionary ideas came directly up against the concepts espoused by the more traditional theories of Justus Leibig. The ensuing give and take between Pasteur's proposal and Leibig's criticisms gave birth to the modern notion of bacteria and other microscopic life. Subsequently, these notions led immediately via logical deduction to the concepts of the "germ" (bacterial) theory of disease and its complement, antiseptic sterlization. I will show you how Joseph Lister (the man Listerine is named after) took Pasteur's theories and made the logical deductions which led him to wash down the walls of Glasgow Hospital's operating room with carbolic acid—which was modern medicine's first successful attempt to render the surgical environment hygenic. Pasteur's case, as I will point out, is a particularly good example of how the discovery phase and the verification phase are closely connected in both logical structure and time. Our final case history will concern an event in contemporary twentieth-century physics. Wolfgang Pauli, an Austrian physicist who has done his major work in America, discovered an explanatory concept—the neutrino—which became extremely successful in tying together some troubled and loose ends in high-energy research. However, although Pauli's discovery occurred in the late 1920s, the acceptance phase of the neutrino theory has strung out all the way until today. Thus, as I will conclude in the last chapter, it is still possible for physicists to ask themselves and one another, "Does the neutrino exist anywhere else but in our minds?"

The range of events involved in these three cases should be a good illustration for you of the two main points I hope to make, first, that all the sciences have similar processes involved in their historical dynamics of discovery and verification, even though, second, the various sciences do in fact really and truly differ from one another in significant respects. Although these are simple points, their conjunction, as I am sure you have already deduced, has the important implication that we must be very careful in our generalizations about scientific activity. However, to say that we will move cautiously in generalizing does not mean that we will not move at all. Science, as I suggested in Chapter 3, can be described in a useful and clear way, even though the resulting description is not universally true in a precise and strict sense for each and every individual science. Thus, although some scientists might want to claim that only precise and strict definitions are acceptable, we can only do the best we can do. And, it seems to me, our best is useful and insightful. With these points in mind, let us now move into a discussion of scientific discovery.

A LOGICAL PUZZLE

Any attempt to combine the discovery/verification view of science with the historical view runs immediately into a logical puzzle. Indeed, this same logical puzzle seems to be involved in any instance of the discovery of new things, whether in ordinary affairs or in science. Probably the best way to introduce the logical puzzle is to state it in the simple and elegant fashion used originally by Plato some twenty-five centuries ago. In Plato's essay "Meno," his spokesman, Socrates, is presented with the following problem.[1] It seems that discovery—that is, reaching new knowledge and information—is not possible. This conclusion is derived from the argument that discovery results from an inquiry either about something we already know, or about something we do not already know. But it is easily seen that discoveries cannot result from inquiries about what we already know, since if we already know something, it cannot be considered to be a discovery. On the other hand, discoveries cannot be results of inquiries about that which we do not know, since if we do not know something, we cannot even inquire what it is we are inquiring about. You can see that this argument depends upon the fact that we can know something only if we can recognize it as being what it is. But we can recognize it only if we already know what it is. This is because if we do not recognize it, then we cannot name it, or describe it, or perceive it, etc., in such a way that our concepts can be applied to it, and so on. To understand this, suppose that something absolutely new comes into our perceptual vicinity. If this something really is absolutely new, then we cannot answer the questions

[1]Plato, "Meno," in R. E. Allen, *Greek Philosophy: Thales to Aristotle*, (New York: The Free Press, 1966), p. 107.

"What is it? What is it like?" Answering these questions would be an essential part of understanding what it is that has been discovered. We can only answer "It is an undescribable something which I have discovered." Admittedly, this is not much of a discovery.

The upshot of this way of arguing, as Socrates himself admits, is a dilemma: Either we cannot ever discover something new, since we could not recognize it in the first place, or every event that we think is a discovery is in fact something we already knew, and thus is not a discovery at all. The ultimate conclusion, of course, is "There is nothing new under the sun." Plato, however, was a wise enough man to recognize that humans were always coming across things that they *think* are new discoveries. Given this psychological fact about human life, he attempted to come up with a hypothesis that explained and made consistent both the logical conclusion that discoveries were impossible, and the psychological conclusion that people often believe that new discoveries have been made. His explanation is a marvelous example of creativity in hypothesizing: He postulated that everyone was born with all the knowledge that he would ever need, but that most of it was hidden in a sort of "unconscious" part of our minds. Here with Plato's hypothesis begins the famous doctrine of innate ideas, that is, the view that humans have concepts which are inborn in their minds. Interestingly enough, the process which we call "education," the process which attempts to bring each of us to knowledge and understanding, is called by a Platonic name drawn from his theory of innate ideas. The word "education" comes from the Latin *educere*, which means "to draw out." Educational processes, thus, consist in "drawing out" from students that which they already possesses in their inborn but unconscious knowledge. The so-called "Socratic" teaching method, which is based on careful question-and-answer dialogue between teacher and student, is modeled on the notion that students are to be carefully led to conclusions which they themselves will make from their own inborn stock of reasoning methods and data. The feeling we sometimes have after a discovery—the feeling "I knew it all the time!"—is quite consistent with Plato's epistemological position as described here.

Although the innate-idea hypothesis is still a very live issue of debate, the logical puzzle "How can I ever discover anything new?" is no longer a main element in this debate. But the puzzle has not really gone away; rather, it has cropped up again (as good philosophical questions always do) but in a different place this time. I speak here of the position we recently considered, namely, that scientific activity involves paradigms, traditions, and revolutions. The difficulty raised for this view by Plato's logical puzzle can be put very succinctly.

It is a straightforward, plain psychological fact that when a person believes "*P* is true" (where *P* is any proposition stating some apparent fact or hypothesis), it is very difficult for that person to accept evidence for the contrary view "*P* is not true." There are many reasons for this fact of human behavior. Obviously included among these reasons are just plain stubborness; and laziness in the face of having to dig out further data which might decide between *P* and ~*P*; and/or fear of being laughed at for believing ~*P* when *P* is a very popular position. But there is another and deeper reason. It has become more and more clear during recent experimental research that our perceptual systems and our cognitive or reasoning systems are not two discrete entities. It has been found, for example, that the sensory organs are not mere extensions of our information-processing brains, but rather that they themselves appear to do some processing all on their own. Thus, eyes and brain, ears and brain, hands and brain, and so on, are merely distinguishable regions of the same system. Each organ is a subsystem of the overall "supercomputer" we call our mind, and all are highly interrelated. Perception/brain interrelations take many forms. For example, as psychological studies have focused more clearly upon the nature of the interactions between perceiving and conceiving, between percepts and concepts, it has become evident that each process counterinfluences the other. Perception determines and influences certain conceptualizations, and certain conceptualizations determine and influence perceptions. It is this second process that I am most interested in hcrc. A main aspect of this process has been referred to as "mental set." According to this notion we often describe patterns of perceptions according to the expectations or prior conceptualizations that we bring with us in our minds into the observational situation. In an extreme example, the paranoid has the mental set to perceive most events as elements in a conspiracy against him. But in a much less extreme example, the implications of this view for science should be getting a bit less murky for you now: Theories and paradigms are conceptualizations which both the individual scientist and the scientific community at large bring with them into the observational situation, into the experimental laboratory. These mental sets must influence the ways in which scientists can experience their data, and must obtrude into their very observations. Given that scientists practice according to paradigms which are central points of traditions—which means that scientists act with a hardened mental set—our Platonic question about discovery is now revived in its new form: If conceptualizations influence observations, and paradigms are conceptualizations, then how can scientists make new discoveries, that is, experience perceptions and observations which run contrary to their mental set of traditional beliefs? This is the guise which Plato's logical puzzle about discovery assumes in our modern times. It is a puzzle which we must dissolve if we ever hope to show how scientific discoveries can be made within traditions. However, before we go on to discuss the scientific cases, let me attempt to familiarize you more concretely with the notion that conceptualizations—mental sets—influence perceptions and observations.

Human perceptual experience, which includes scientific empirical observation, is a complex but unified set of phenomena. However, even though this complex entity is a unified whole, it is still possible to *distinguish* between two different elements of this whole. I do not mean that the two elements ever appear separated, or apart from, one another. Rather, perceptual experience always includes both elements simultaneously. But even though we never experience the two distinguishable elements of perceptual experience separately, we must conclude that they exist on the basis of analysis and logical reasoning about the nature of our experience. The two elements I am talking about are (1) the sensory element of perception and (2) the pattern, or arrangement, of the sensory elements. The sensory element may be understood to be the basic "raw qualities" such as particular sights, sounds, tastes, feels, etc., which are the output of our sensory organs. Examples of these would include raw qualities such as a particular color of such and such a hue and intensity; or a sound of a particular pitch and intensity; or a particular feeling of wetness; and so on. The ordinary view (which I accept here) is that these raw perceptual qualities, the sensations in our experience, correspond in some incredibly complex fashion to the properties of objects. For example, the particular sensation of wetness I am experiencing right now might be taken to correspond to the moisture condensed upon the surface of the beer glass I am holding. Similarly, the particular golden color I am experiencing right now might be taken to correspond to the color of the beer. In general it is taken that there is a correspondence between the real, metaphysical features of the world (the properties of objects) and the epistemological features of the world (the raw qualities of our perceptions).

But our perceptual experience does not consist merely in a disparate collection of unrelated raw sensory qualities. Sensory qualities always occur in patterns or arrangements or interrelations. The simplest example of a pattern element in our perceptual experience is the relation we call *temporal sequence*. Temporal sequence (or "time series") is probably our most fundamental perceptual pattern. All raw qualities are experienced in a network of temporal relations; accordingly, some qualities are experienced as coming "before" others, others are experienced as coming "after" still others, while some, of course, are experienced as occurring "simultaneously with"—at the same time as—others. It is pretty clear that there is no metaphysical property of the world which corresponds to the experience "before" or the experience "after." Thus, although we can locate the metaphysical property of an object which corresponds to our experience of the raw quality "red," we cannot find any metaphysical property of any object or event which corresponds to the experienced pattern "before" or "after." A sound might be "loud," "musical," "low-pitched," but it is not "before." This unusual result was discovered by the German philosopher Immanual Kant.[2] Since his time (the late 1700s) there has been general, albeit sometimes grudging, agreement among philosophers and psychologists that the time-series pattern, the sequential temporal arrangement of our perceptual qualities, that we call our experience of "time," in fact is a pattern according to which raw perceptual qualities are arranged, and it is not itself a raw perceptual quality which corresponds to some specific metaphysical property of the world.[3]

The obvious question which bursts upon us here is: If temporal patterning is not provided by our perceptual organs responding to the world, then where does this patterning come from? We can account for qualities such as redness, coldness, bitterness, loudness, etc., all our raw perceptual qualities, by reference to real properties of the world. But we cannot do that for patterning such as temporal patterning (among others such as spatial patterning and causal patterning). Where, then, does this patterning come from? Kant's answer is that our cognitive system itself provides the pattern. Raw qualities arc always arranged in such a way that we experience them as occurring before, or after, or simultaneously with one another. Thus, the temporal patterns we discover in our perceptual experience are part of our basic mental response to the world; in a word, the temporal pattern of the world is a fundamental and basic sort of mental set which we bring with us to our first experiencings of the world. To give another extreme example, just as paranoids arrange their raw perceptual qualities in such a way that they experience them as threatening, so, in all cases, humans experience their raw perceptual qualities as temporally patterned. Let me give you another example of a fundamental and automatic patterning contributed by our cognitive systems. But first, a word about the language I am using.

[2]Immanuel Kant, "Transcendental Aesthetic," in *Critique of Pure Reason* (New York: St. Martin's Press, 1929).

[3]An apple as perceived, for example, is a patterned arrangement of color, taste, consistency, etc., qualities. Each of these qualities corresponds to some property of "real" apple, i.e., where "real" means the physical apple as distinguished from my perceptual experience of it. But, if I experience my perception of the apple *after* I have my experience of the apple tree, I cannot say that there are *three* real features of the world, namely, "apple tree," "after," and "apple" which correspond to the sequence "apple experience came after apple tree experience." That is, while "apple tree experience" corresponds to some physical entity, and "apple experience" correlates to some physical entity, it is not so clear that "after" similarly corresponds to some discrete physical entity.

If raw perceptual qualities of our experience are patterned by the mind itself, then it is clear that, were the pattern to change, the character of our experience would change. I use the odd expression "pattern experienced *as* an . . . (*x* or *y* or whatever)" to indicate this. The "as an . . ." phrase indicates that the raw perceptual qualities are experienced to *be* something, let us say, to *be* an *x*, but that they could perhaps have been experienced otherwise—"as a *y*" instead, for example. You will see what I mean in just a moment.

The psychologist T. G. R. Bower has done some fascinating experiments with very young babies—some less than three weeks old.[4] He has shown that certain patterns of experiencing are apparently imposed by babies upon their raw perceptual qualities even at such young and tender ages. As Bower himself notes, he set up his experiments in order to try to get information about this question which had been suggested by many philosophers, including Plato, Augustine, Descartes, and Kant. In one experiment, Bower set up a slide projector in such a way that it projected a three-dimensional visual image very near to a baby's body, where the infant could both see it and reach it. The image was of a brightly colored, solid, ball-like object. The babies quite naturally reached for the image. When their hands went through the apparent region of the ball without encountering anything that gave any solid "feels," that is, any raw tactual qualities, the babies exhibited a startled response, and got quite upset. The interpretation of this experiment is actually quite straightforward. The infants experienced the raw visual perceptual qualities *as a three-dimensional physical object.* That is, they arranged the *visual* data alone in such a way that they produced the experience of a solid, physical object. But part of the expected *total* pattern of a solid, physical object is that it will produce raw tactual feel qualities if one reaches out and touches it. But the apparent ball-like object produced no feels in the babies' hands. Thus the induced pattern (that is, the babies' mental set or expectation that visual input such as they were having would be accompanied by an experience of the pattern of a solid, three-dimensional physical object) was wrong, and this mistake upset the babies.

Bower gives many examples of how babies pattern their perceptual experiences, of how they make certain specific perceptual qualities conform to certain patterns. Basically, it is pretty easy to see the probable reason behind what is going on here. Human beings operate in a world in which many objects are of permanent types, but these objects move around and change rapidly, producing in their human perceivers only sparse and fleeting pieces of raw perceptual qualities. We thus are

[4] T. G. R. Bower, "The Visual World of Infants," *Scientific American*, December 1966, p. 80; "The Object in the World of the Infant," *Scientific American*, October 1971, p. 30.

required to make rapid and correct perceptions on the basis of scarce and hurried sensations. In order to produce an informative totality out of this scarcity of data, it has been necessary for the mind to develop the ability to set the sparse data into useful patterns. Thus, we have come to have expectations, or mental sets, about how to organize and produce meaningful patterns out of the impoverished and fleeting sensual data. It is easy to suspect that the long eons of evolution have helped to select those persons with the most valuable mental expectational patterns. For example, those early ancestors of ours whose mental set patterned the fleeting visual qualities "moving" and "yellow-and-black striped" into the experience "pouncing saber-toothed tiger" probably had a better chance of surviving long enough to leave descendants than did their colleagues who patterned this same set of raw qualities into the experience "falling, overripe banana." Let me give you one final example of how our experience consists of raw qualities patterned *as something.*

The British psychologist Richard Gregory has done quite a bit of experimental work on the relation between raw perceptual qualities and how we pattern them.[5] He has hypothesized that significant aspects of the patterns of various perceptual experiences are in fact contributed by the perceptual system and brain system working together. He has concentrated especially upon those paradoxical figures so gleefully displayed in all introductory psychology books. The Necker cube (see Fig. 4-1), the reversing staircase, the duck/rabbit picture, and the birdbath/two-faces figure are all explicit examples of perceptual "underdetermination," i.e., cases where the metaphysical properties of the object are too sparse to allow complete determination of which perceptual pattern is required for fit. Gregory argues that the perceptual qualities of these figures are not sufficient to fix a pattern strictly and precisely in our perceptual experience. Because of this, the mind vacillates between two possible patterns or arrangements for the raw perceptual qualities; sometimes we perceive the figure *as a cube pointed our way*, but then, with absolutely *no change*

[5]Richard L. Gregory, "Visual Illusions," *Scientific American*, November 1968, p. 66.

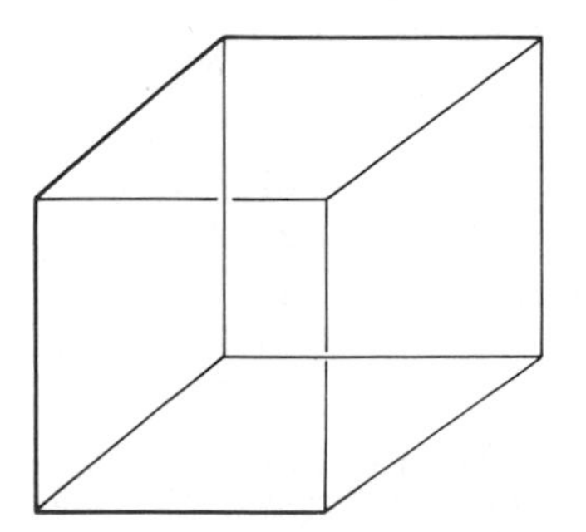

Figure 4-1. The Necker cube.

in the qualities of the object involved, we perceive it *as a cube pointed the other way*. Gregory believes that human evolutionary history has prepared our cognitive systems for only a limited range of perceptual qualities, and that thus there exists only a limited number of various patterns, various ways, of perceiving the sets of qualities.

Until this point, I have only provided examples in which the patterning is provided by unconscious mental mechanisms which appear to be inborn and automatic features of *Homo sapiens* as a species. However, the idea that the mind provides patterns, mental sets, which structure and organize the raw perceptual qualities that we take in from the world, is not limited to these primitive and automatic situations. Similar sorts of processes can apparently be extended to include cases involving material we would think of as being much more conscious and abstract. But before getting down to the details of more refined particulars, consider a simple introductory case of conscious mental set.

Suppose that your neighborhood has been the target of a very successful series of nighttime burglaries. These dark intrusions usually have occurred via back-of-the-house window entries in the early hours of the morning. You have been reading newspaper accounts and watching the TV news about these burglaries for several weeks, and you are understandably keyed up about the whole thing. You go to bed each night a little bit more worried. Tonight, all of a sudden, you wake up at three o'clock and hear the burglar in the back bedroom. You dash in, six-gun in hand (or more prudently, quietly call the police and let *them* dash in, six-guns in hands), only to discover that the sound which you perceived *as the burglar* in fact should have been perceived *as an element in the pattern of the house settling on the foundation*. Thus, you perceived the creaking board *as the burglar* instead of *as the house settling*. In this example, an item of conscious experience—the fear of burglaries—has provided a ready-made pattern, an expectation or mental set, within which you placed the raw perceptual quality "creaking sound." Conscious adoption of patterning elements has been the subject of quite a bit of investigation.

The American psychologist and linguist Benjamin Whorf noted that specific linguistic communities, that is, specific populations which possessed certain delimited groups of concepts expressed by their language, apparently patterned the world differently from other linguistic communities.[6] Eskimos, for example, have a large number of concepts for snow. Thus, they perceive as patterned among the world's properties many more distinct types of snow than does a community living in the tropics. For the Eskimos, raw perceptual qualities are patterned *as snow type 1*, or

[6]Benjamin Whorf, *Language, Thought, and Reality* (Cambridge: Massachusetts Institute of Technology Press, 1956).

as snow type 2, and so on, in a fashion quite different than could be expected from say, the Navaho Indians.

The American philosophers of science N. R. Hanson[7] and Thomas Kuhn have each suggested versions of the general view described here. Hanson, in one instance, asks us to consider the pattern which an eighteenth-century American would impose upon the perceptual qualities produced by an electric light bulb. It is unreasonable to suppose that our hypothetical observer would pattern the object in any other way than *as a particular glass-brass object*. It is epistemolgically impossible (*and* historically impossible, I might add) that he could have patterned it *as an electric light bulb*. In a different example from Hanson's, Kuhn points out that sixteenth-century astronomers believed that the sun went around the earth. Thus the perceptual qualities produced by watching the eastern horizon at dawn are seen by the sixteenth-century observer as sunrise, as the rising movement of the sun higher in the sky. But contemporary, twentieth-century astronomers cannot provide this pattern. They experience the qualities of the situation as *earthturn*, as the rotation of the earth within the light-exposure cone of the sun's radiation.

The ultimate implication of this view for the concept of scientific observation should not be too difficult to see. Since theories and paradigms function as conceptualizations and mental sets which pattern raw perceptual qualities, and since the raw perceptual qualities are so scarce as to underdetermine the range of patterns applicable, then theories and paradigms are highly significant elements in what scientists experience as their observations. That is, the perceptual experience, or perceptual world, of the scientist is observed as patterned by the concepts of the respective conceptual structure existing in their minds. Here and now Plato's puzzle comes suddenly into view, looming large in our normal idea of science: If scientists can experience the world only as they know it, in other words, if they can observe the raw perceptual qualities solely as patterned by their concept structures—the theories and paradigms which are currently functioning in their tradition—then how can they ever come to see the perceptual qualities patterned *as something new, as something unknown, as something outside the range of patterns specified by their traditional concept structures*? To put it most succinctly, how can discoveries be made? This is the central question I wish to answer. Accordingly, I will attempt to show how concept structures such as theories and paradigms function in their clash with new experiences or with new patterns.

But before I go any further, I want to issue a warning word or two.

[7]N. R. Hanson, *Patterns of Discovery* (Cambridge: Cambridge University Press, 1965), pp. 6ff.

Kuhn and others, Whorf included, have been criticized for overstating the case about the influence of concepts upon the perceptual situation. Kuhn and Whorf, in particular, seem at times to want to claim that people who have concept *x* live in a world which is actually, metaphysically, different from people who do not have concept *x*. A Christian, for example, experiences the world as having a certain pattern which is not perceived by the atheist. But beyond this literal sense it seems unsafe to go. It is quite all right to say that the Christian experiences the world *as being different* than does the atheist. To go further, however, and say that simply because his *experience* of the world is different, the world of the Christian really is different, is to go too far. Basically this criticism denies the idea that the world contains object *x* simply because we experience it *as containing x*.

It seems to me that the truth of the matter lies somewhere between the extreme view that the world is just as we experience it to be, i.e., just as we pattern it to be (this is the view that all paranoids are correct), and the other extreme view that we experience the world just exactly the way it is, nothing more, nothing less (this is the view that the world must contain properties like "before" and "after," and so on, in addition to properties like "red" and "square"). To put it in very succinct terms, but not necessarily the most explanatory terms, scientists (like all the rest of us), tend to see the world just as their mental sets tell them to, until and unless some property of the world intrudes so forcefully upon them that they are deflected out of their mental set. Discoveries in science, it seems to me, result from just such deflections. The next three chapters will give you some examples of this process at work.

SUGGESTIONS FOR FURTHER READING

Two very different accounts of discovery can be found in N. R. Hanson, *Perception and Discovery* (San Francisco: Freeman, Cooper & Company, 1969); and Stephen Toulmin, *The Philosophy of Science* (N.Y.: Harper Torchbooks, 1960). A third view can also be found in Sir Karl Popper's classic *The Logic of Scientific Discovery* (N.Y.: Harper Torchbooks, 1968). Popper also discusses his own views on verification in this book. A book I like because of its common sense, its actual science feel, is John Ziman's *Public Knowledge* (Cambridge: Cambridge University Press, 1968). Ziman has some very good things to say about the publicly demonstrative nature of verifying theories. And he says it in plain language. On the question of the role of theory in observation, Hanson's books are good places to look. I am also able to recommend one of my own pieces in this controversy: Edward Walter and George Gale, "Kordig and the Theory-ladenness of Observation," *Philosophy of Science*, Sept. 1973, p. 415. As far as discussions of Kant's theories are concerned, it is tough to find any which can be recommended to the nonspecialist. However, I think that one of the best is T. E. Wilkerson, *Kant's Critique of Pure Reason* (Oxford: Clarendon Press, 1976). See pp. 54–57 especially.

Chapter 5

Lavoisier and Oxygen

INTRODUCTION

Modern chemical science has just celebrated its bicentennial. Chemistry, as least in its modern format, is usually dated from Easter 1775, when Antoine Lavoisier published his memoir on combustion and calcination.[1] A minority of historians (which includes me), however, date the founding of modern chemistry from two years later, August 1777, when Lavoisier published the revised and corrected version of the Easter memoir. But even given this quibble over dates, it is generally accepted that modern chemistry originated sometime during this two-year interval because of a tremendously significant discovery made by the young Frenchman Lavoisier. It will be my task in what follows to acquaint you with this story. I hope that you will find it entertaining, as well as enlightening in regard to scientific discovery.

[1]James Bryant Conant (ed.), *The Overthrow of the Phlogiston Theory* (Cambridge: Harvard University Press, 1950). This pamphlet is Case 2 in the Harvard Case Studies in Experimental Science. It is an extremely useful little book, and I shall rely heavily upon it in my discussion. You are encouraged to get yourself a copy; it is cheap, and supplies a lot of data I am not going to be able to include.

But first let me say a couple of very general things about discovery. As I pointed out in the previous discussion, all discoveries take place within a tradition defined by a paradigm, that is, they take place against a backdrop provided by the prevailing mental set. However, there are discoveries and there are *discoveries*. Some discoveries in science are not terribly significant or disconcerting. For example, adding another decimal place to the measured speed of light is not terribly earth-shattering. For another example, if it had been thought for a long time that all crows were black, and that was the mental set, then it would not be terribly important if a white crow were found. Some explanation, e.g., mutation, would patch up the general concept about the color of crows. Similarly, if it were thought that all falling objects near the face of the earth accelerated at 32 ft/sec^2, then it would not cause much consternation to find an object which accelerated at only 31 ft/sec^2. Closer measurement would probably show excess air friction or some other such anomaly. What I am getting at here is that science is always marching on, at least in the sense that it gets more precise values for common measurements, and finds individual exceptions as it makes ever closer scrutiny of limited domains of study. Discoveries of this sort are not overwhelming in any way.

Other discoveries, however, are of a type that is extremely significant. For example, suppose that a class of physical objects, objects otherwise physically normal, were discovered which did not react positively to gravity. Obviously, these objects would behave quite unexpectedly in gravitational fields. For example, they would not accelerate at 32 ft/sec^2 when dropped; indeed, they would not accelerate at all. In the face of this kind of discovery, we could not fix up the theory by merely fiddling with the values of the numbers and the connection between secondary concepts. Rather, we would be faced with such a massive incongruity between our normal ideas and our newly found experience that we would be forced to completely rethink our position in regard to gravity, acceleration, falling bodies, and so on. Such a vast rethinking in our theories is called a "revolution" in a paradigm. It is just such a vast project that Lavoisier was engaged in. After Lavoisier's work, a revolutionary new chemical paradigm had been created. This new paradigm was not simply a patched-up version of the old one. It was radically different, as I shall make clear. As I noted in the earlier section, my main problem in explaining and describing this case will be to show how Lavoisier's discovery resulted from a collision between his beliefs in the older theory, beliefs which shaped his observations of chemical phenomena, and some real occurrences in the physical world. In general, this will be the problem I will face in dealing with all cases of revolutionary discovery.

In the main, I will give an account of Lavoisier which will have a logical structure applicable to all other cases of revolutionary discovery.

Let me state it now in broad terms. My view is that there is an interdependence between what scientists believe and what they see, that is, between their mental set and what they experience as their observations. Moreover, this interdependence exists in whatever domain they are operating, whether it be in strict science, or everyday affairs or whatever. We have already seen that this view is dangerous in that it might logically be extended to assert "All paranoids view the world correctly." But I do not think that we have to go that far. Rather, what I believe is that each of us, including Lavoisier, believes at any one time a multitude of theories; thus, each of us has a myriad of mental sets which we carry with us into the laboratory of experience. And each of these mental sets generates perceptual expectations. Consequently, it is plausible to suppose that sometimes there is logical inconsistency between the expectations we have about particular objects, this inconsistency being due simply to the fact that more than one theory is being brought to bear upon that particular object. Discoveries, it seems to me on this view, are the result of attempts to make our various particular mental sets, our various theoretical expectations about the world, logically consistent with one another.

Lavoisier is a prime example. Lavoisier began his combustion experiments from within the exact same mental framework as all his compatriots, at least so far as chemical theory is concerned. He and all his colleagues learned phlogiston theory in school, and thus it was phlogiston theory that they applied in their laboratories. But this brings up a sensible question: Given that Lavoisier had been taught and thus believed the same concepts as his colleagues, why did *he* make the revolutionary discoveries, and not any of the others? This sensible question has an equally sensible answer, in terms of the theory I am here espousing: Lavoisier must have had some particular specific theory (or theories), his own peculiar idiosyncratic concepts, *in addition* to the phlogiston theory, which generated mental sets *in addition* to those generated by phlogiston theory. He believed *all* these theories even as he began to do his experimentation. On this account, then, what needs to be looked for is a specific belief (or beliefs) which might have forced Lavoisier to experience the situation, that is, to make his observations, in a fashion different from his contemporaries. Is there such a set of concepts? Obviously (or I would not be writing this) there is. In fact, there are two separate elements. The first one is this: Lavoisier had had some training in law and business; he knew full well what it was to quantify information, and to keep balanced accounts and ledger sheets of these quantities. Thus, he had a certain mental set, a psychological attitude if you will, which tended to set him up to perceive the world in categories which might be quantified. In a word, he brought a penchant for accountancy into his chemical laboratory. However, in addition to this psychological element, Lavoisier also brought a particular piece of scientific knowledge: He knew well what was going on in the physics of his time, and it was the extension of application of his understanding of contemporary physics into chemical affairs which ultimately led to his complete reconstitution of the earlier phlogiston paradigm and its data.

Given this view, my account of Lavoisier's discovery will have three things to make clear. First, I must describe the prevailing paradigm, the overall theory which Lavoisier and his fellow chemists were taught. Then I must indicate how Lavoisier's business mentality interposed a particular pattern into the paradigm. Finally, I must show how Lavoisier's understanding of physics was brought to bear. Let us now start with Lavoisier's original chemical paradigm, phlogiston theory.

CONCEPTUAL HERITAGE

Phlogiston theory, which I briefly mentioned in Chapter 3, was the general theory which chemists were taught during the late eighteenth century. Phlogiston theory was the ultimate outcome of the Aristotelian paradigm as it applied to chemical matters. The Aristotelian paradigm, plus its modifications during the alchemical period, had culminated in the view that air, earth, fire, and water were the fundamental constituents of all chemical substances. Phlogiston itself was conceived to be a physical manifestation of the element fire as it was involved in combustion and smelting. However, before we get to the precise details of phlogiston theory, let me briefly mention some of the peripheral modifications which had been made in the general Aristotelian chemical paradigm between the time of the alchemists and Lavoisier's own era.

During the century immediately prior to Lavoisier's work some modifications had been made in the general Aristotelian conceptual system as it applied to chemistry, and these modifications in part prepared Lavoisier's generation to expect further modifications. One line of research had focused upon the Aristotelian element air. Scientists such as Von Helmont, Black, and Priestley had done experiments which apparently showed that air, which was a fundamental and simple element according to the Aristotelian view, on the contrary might not be an element in itself; that is, the experiments apparently showed that air might be a compound substance rather than a pure elementary substance. Additionally, Black had isolated an "air" which was different from ordinary, garden-variety air. From these experiments a general but rather fuzzy notion of a "gas" was floating around in the minds of Lavoisier and his chemical colleagues. These sorts of results had at least produced some expectation of future changes in the basic Aristotelian theory. However, in and for themselves they did not constitute discoveries, or even significant data to be used in any major overhaul of the Aristotelian chemical theory. It was to be Lavoisier's work with phlogiston theory which finally made sense of all these earlier researches, it was his work which tied them together into a unified context of theory plus observable correlations. The main thing to note about all this is that chemistry had entered into a period of ferment, not unlike that which had befallen physics a century earlier during the period before Newton, that is, around 1660–1680. In fact, analogy between the roles of Lavoisier and Newton is significant in the aspect that both men's work produced the keystone which suddenly synthesized and made coherent and interlinked the diverse results of earlier researches. However, we ought not to push this analogy too far at the moment. Let us instead examine the intellectual apparatus—phlogiston theory—which Lavoisier acquired in his studies and early researches.

PHLOGISTON THEORY

Phlogiston theory had been developed as a hypothesis to explain the details of smelting, combustion, and calcination. (*Calcination* is the process of turning a metal into an earth—the rusting of iron is a typical example. If you rub the rust off a piece of iron, it is clear that the nature of the metal in the rust has been transformed, and the metal has become a powdery, earthy substance. We now have a different name than "calcination" for this process, a name which you probably already know, but the modern name did not come into use until *after* Lavoisier's work. Hence, I cannot use the modern name until *after* I tell about Lavoisier's discovery.) The observable correlations involved in smelting and other aspects of metallurgy had been known for some time, and indeed, knowledge of some phases of metallurgy—iron production in particular—had been around since the time of the Greeks. But a big boost in metallurgical information had come when Agricola (1490–1555) published his work on mining, smelting, and related metallurgical techniques.[2] His book was a compendium of all that was then known about the arts of producing and refining metals. But, even given its technological significance, it is quite safe to say that Agricola's book gave descriptions which were strictly techniques; indeed, the book was nothing more than a cookbook. It consisted of empirical observations correlated together into recipes which could be used to produce the desired metallic end product. The observational correlations reported in this book, together with later additions,

[2]Agricola, *De re Metellica* (Basel: 1556).

were to become the data linked to the concept structure of phlogiston theory.

Let me briefly recapitulate my notion of scientific theory. As you will remember, my view is that a scientific theory roughly consists of two parts: First, a set of purely empirical "if, then" conditional correlations (remember the Egyptian astronomical algorithms) sufficient to allow reliable prediction of the outcome of observational events; second, a set of concepts (remember the Greek metaphysical ideas involving concentric, rotating spheres containing the embedded heavenly bodies) sufficient to explain the correlational scheme causally. In respect to this view of scientific theories, metallurgical recipes were fairly simple things. One might find, for example, reference to "so and so many buckets of such and such an iron ore," "such and such an amount of charcoal," and finally, "a fire heated to some particular heat"—this latter instruction being given in terms of the color of the fire.

Empirical Observations

A beautiful example of this sort of empirical recipe, one which additionally is preserved as part of a religious ritual (a not uncommon means of preserving empirical knowledge) is found in J. Brownowski's depiction of the Japanese swordmaker in *The Ascent of Man*.[3] The ritual very precisely and explicitly calls out each and every step of the procedure, down to the exact shade of the fire's color. Illustration of the difference between the empirical and the theoretical aspects of science is provided by the divergence between the ritual's merely telling *how* to produce the sword, and Bronowski's running commentary (plus microphotographs) explaining *why* each step contributes to the ultimate internal microstructure of the finished sword. I doubt whether Bronowski himself could have made the sword; but he surely does tell why the master craftsman proceeded in exactly the fashion he did. Let us take the following to be an example of the sort of recipe found in metallurgical texts:

> Take 2 parts iron ore, add to three parts charcoal, mix and heat over a fire kept burning with the color of the midmorning sun. This will produce four parts of soft iron metal.

In order to make this recipe a little easier to deal with, I will now take the liberty of expressing it in a somewhat more modern and symbolic form, rather akin to an equation. Even though this type of representation did not

[3]J. Bronowski, *The Ascent of Man* (Boston: Little, Brown and Company, 1973). The account of the Japanese Swordmaker is found in chap. 4, "The Hidden Structure."

come into use until sometime after Lavoisier, it will make the flow of thought clearer for us. (Additionally, although my use of an equationlike representation is a bit unhistorical, I will soon point out that the idea of an equation, an identity between ingoing ingredients and outcoming products, is extremely significant in Lavoisier's thought.) Here is the above metallurgical recipe expressed in an equation:

$$2 \text{ parts iron ore} + 3 \text{ parts charcoal} \Rightarrow 4 \text{ parts iron}$$

As it stands, this recipe and others like it are simply nothing more than correlations expressing the observations which have been made over a long period of time. They are simple "if, then" conditionals which, in and of themselves, do not claim any causal relations between ingredients. It remains for some conceptual system to explain how and why the correlations are as they are. In this case, phlogiston theory became the first candidate for the status of "explanation" relative to the metallurgical recipes.

Phlogiston Concepts

Prior to describing the details of phlogiston theory, I must utter a word of caution. Phlogiston theory is a significant scientific and intellectual creation. But it will most likely sound a bit strange to you, and because chemical theory has markedly changed, it may sound silly and even a bit stupid. However, the creators of the theory, and scientists such as Lavoisier and Priestley who later came to study and use it, were making an obvious leap in progress as compared to those who earlier had rested content with mere empirical observations and cookbook metallurgy. Any attempt, successful or not, to pass beyond observable correlations into some explanatory scientific scheme is a risky endeavor. It takes a certain amount of bravery to postulate unobservable objects in order that they might explain observable events; and it takes skill and intelligence to postulate *well.* On these criteria, the inventors and proponents of phlogiston theory deserve our respect, even though on other grounds we might feel embarrassment at the oversimplicity and, indeed, ignorance embodied in their conceptualizations. Embarrassment such as I mention here is perhaps the main reason why most modern chemistry textbooks cover up phlogiston theory. Nothing, however, need be covered up. Phlogiston theorizing was an honest attempt, made by dedicated and intelligent men, to make some sense out of the vast diversity of facts presented by the recipes. Because we now think the theory wrong brings no criticism to bear upon the skill and scientific acumen of its creators. With this now in mind, let me herewith begin to lay out some of the details of the phlogiston conceptual system.

In the Aristotelian paradigm, as we saw earlier, the substances earth, air, fire, and water were fundamental. Associated with each of these substances were similar qualities, which grew into parameters for classification of naturally occurring substances. Thus, ores, clays, rocks, and so on were conceived to be members of the "earthy" category of substances. Water, oils, milk, wine, and so on were taken to be "watery" stuffs. "Earthy," "watery," "fiery," and "airy" were qualitative descriptions which could in principle be applied to most of the substances men commonly encountered. On this view of the world, it is plain to see that the chemically inclined scientist would come to view qualitative explanation as the objective of science. That is, what became necessary to be explained were the observable qualities of individual kinds of things. For example, the scientist might attempt to account for the shininess of some particular sort of metal, or the slipperiness of some particular liquid, or even the density or lightness of some other kind of special stuff. Along these lines, iron ore came to be seen as an earthy sort of substance known as a "calx" (calxes were produced by the "calcination" process). Charcoal was also earthy to an extent, but even more significantly, it was mainly "fiery" for the obvious reason that it burned freely and cleanly, and left only a small amount of earthy residue in its ash.

By at least thirty years before Lavoisier the criteria implicit in the above description had been made explicit: Concepts postulated to explain observations had to satisfy the demands involved in a qualitative explanation. That is, they had to account for the initial qualities of the ingredients of a chemical reaction, as well as any qualitative changes which occurred during the course of the reaction itself. Thus, the chemistry studied by Lavoisier focused upon qualities, rather than upon the quantities involved.

Further, the qualities involved were overall related into a scheme handed down from Aristotle and the alchemists: Airs, earths, fires, and waters were the ultimate qualitative categories into which substances were arranged. Phlogiston was the substance which was either fire itself, or some fundamental manifestation of the fire principle. (Chemists argued among themselves about which of these interpretations was correct. The more cautious among them chose the latter view.) According to the chemical paradigm, individual substances were usually compounds of the four basic types, and transformation of particular substances into other categories was thought to be one object of chemical manipulation.

But phlogiston theory was not a purely Aristotelian development. In the century preceding its acceptance in chemistry, physics had undergone a swift, wide-ranging, and successful conceptual revolution. The theories of Copernicus, Kepler, Galileo, Descartes, Huygens, and others had become suddenly interknit by the work of Newton. More importantly for our story, these successes in science had occurred within the framework of an atomistic ("corpuscularian") metaphysics, as I noted in Chapter 1. The atomist view that matter was composed of unobservable, hard, massy, and moving particles had thus become dominant in physics. Because of this metaphysical belief, physicists concerned themselves with interpreting the observable world in terms of the masses and motions of these unobservable corpuscles. Some carry-over of these ideas into chemistry would be expected.

Using the atomist metaphysical scheme, Robert Boyle had defined two notions which were later to become crucial in the chemical revolution. Boyle, a contemporary of Leibniz and Newton, and most famous for Boyle's law of gases, had proposed that material substances existed in three states: solid, liquid, and "air."[4] Each of the states was dependent upon the relative density of the underlying atoms. "Airs" thus are diffuse collections of the small unobservable corpuscules. Liquids and solids are, respectively, each a more compressed state of the former. Boyle's notion here is an inherently quantitative one, one which explains qualitative states in terms of quantities of objects distributed per quantity of spatial region. It is most likely because of this emphasis on quantitative analysis that Boyle's idea of the three states of matter did not have an immediate influence in chemistry. A qualitative science such as alchemistry obviously would have trouble utilizing a quantitative scheme like the one he proposed. Only later did the new chemistry, a revolutionarily new *quantitative* chemistry, come to make use of Boyle's idea.

Boyle's other conception also did not have direct and immediate impact upon his contemporary chemistry, although it did come into great significance a century later during Lavoisier's ascendancy. Boyle's general idea was to define the chemical atom as "that which could not be further reduced into components via chemical means." However, although this idea did not itself have a direct impact, it did produce an effect through an attendant ramification. His original idea implied a notion of chemical "analysis," that is, the decomposition or dissociation of a compound substance into its simpler and more elementary parts. This idea that ordinary everyday observational substances were merely gross manifestations of underlying collections of atoms/corpuscles took strong hold in the minds of the chemists. Thus, even though the specific idea of

[4]Boyle's work on the "airs" has two aspects, chemical and physical. Boyle's general physical notion is that air is an "elastic fluid," that is, a fluid which, unlike fluids such as water, has "spring" and thus may be compressed. See *Robert Boyle's Experiments in Pneumatics*, James Bryant Conant (ed.), in the Harvard Case Histories Series (Cambridge: Harvard University Press, 1965). In regard to their chemical aspects, however, Boyle's ideas do not offer much beyond the idea that the various "airs," such as the "fixed air" given off during combustion and respiration, are varieties of the everyday garden-variety air that we breathe.

the chemical atomic element did not take hold, the corresponding idea that atoms constituted all objects did. On this view chemical reactions soon were understood to involve the rearrangement or movement of smaller, unobservable, constituent particles. Accordingly, reactions could be of only two basic types: movement toward each other (combination) and movement away from each other (dissociation). All explanations were to be given in terms of these two.

APPLIED PHLOGISTON THEORY

Given this system of explanation, the task of the chemist was relatively clear. He was first to catalogue his careful observations of various reactions. Then, using these correlational schemes, he was to attempt to account for the observed chemical changes in terms of the qualities involved, and the movements of underlying substances which were the bearers of the qualities. Phlogiston theory fit very nicely into this logical and epistemological framework. Phlogiston was believed to have various properties which explained the qualities observed in the substances of which it was part. As mentioned earlier, phlogiston's most essential property was its fiery quality, which produced the obvious observable effect we call "flammability." That is, a substance rich in phlogiston is very flammable: It catches fire easily, burns cleanly, and leaves little residue. Burning, the process itself, involves the liberation of phlogiston from the flaming substance. Consider the charcoal's burning, for example. Charcoal seems to be a combination of phlogiston, some particular earthy substance, and water. This analysis can be reached simply by noting three points: (1) Charcoal burns, hence it contains phlogiston; (2) charcoal's combustion leaves a small earthy (ashy) residue, and consequently it must contain a certain amount of earthy substance; (3) finally, charcoal's burning liberates water, as can be easily demonstrated by collecting it on a cooled surface held above the burning coals.

Combustion

Combustion, according to this example, is a decomposition reaction. Charcoal's combination matrix of phlogiston, earth, and water breaks down during burning, with the result that its three constituent elements now exist separately, rather than in combination. The diagram in Fig. 5-1 illustrates this in a pictorial fashion.

As the diagram indicates, phlogiston itself is liberated into the surrounding atmosphere. This produces observable effects. For example, air, when combined with excess amounts of phlogiston, becomes "fixed"; that is, it can no longer support "alterations" such as further combustion, or animal respiration (which is akin to combustion). The explanation of

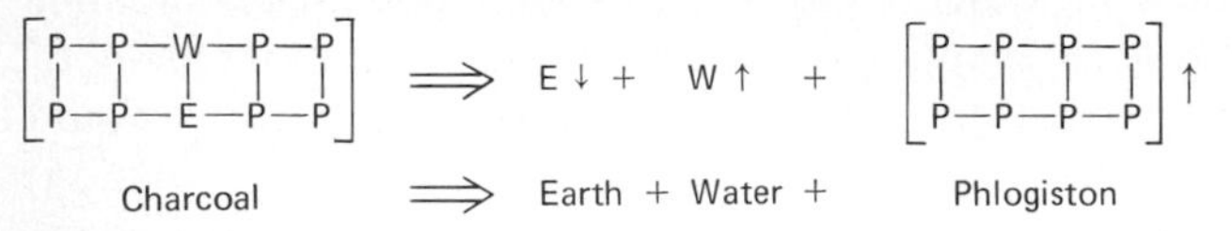

Figure 5-1. Hypothetical picture of charcoal combustion. Charcoal is a combination of phlogiston (P), an earthy substance (E), and water (W). During combustion, the three elements dissociate, with the consequent liberation of water vapor and phlogiston particles.

this fixity, or stable state, is quite straightforward. Everyone is familiar with the phenomenon of "oversaturation." Under normal conditions, many substances can go into solution in others; for example, sugar can dissolve in water. However, there is a point beyond which no further dissolution occurs. Thus, no matter how long and hard you stir your iced tea, that last little bit of sugar just will not dissolve. The tea has become saturated with sugar, and just will not accept any more. Air and phlogiston exhibit a similar relation. Beyond a certain point, air simply will not accept any further phlogiston; it is at that point saturated and fixed.

Professor Samuel Williams, Hollis Professor of Mathematics and Natural Philosophy at Harvard (1780–1788), has provided a convincing demonstration of this fact.[5] His apparatus and method are quite plain and simple: He starts with a bell jar of reasonable size, filled as normal with plain, garden-variety air. Into this bell jar, he plunges a burning piece of paper. If we repeat the demonstration, we can easily observe the results. We will soon see the fire diminish, waver, and finally snuff out. Clearly, the restricted volume of air inside the bell jar has become saturated with phlogiston, and will no longer accept any more into solution. Thus, the liberation of phlogiston from the paper halts, and since combustion is nothing other than the liberation of phlogiston, burning stops as well. This explanation has further testable consequences. Suppose that we increase the volume of the bell jar. The implication is that this larger volume of air will allow a longer period of combustion. Subsequent observation verifies the hypothesis.

In addition to these observable effects, fixed air itself can be detected. If an open container of limewater—water which has a small amount of lime dissolved in it—is placed in the bell jar following the burning, the limewater will very soon turn from clear to cloudy white. This test is definitive, and is itself the key to the earlier discovery that animal (and human) respiration produces fixed air. Joseph Black, among others, conclusively demonstrated in 1764 that limewater turns cloudy

[5]Conant, op. cit., p. 15.

upon exposure to human exhalation. It was on the basis of this evidence that it was concluded that animal respiration and combustion are akin to one another, since both produce fixed air.

Dissociation, then, one of the two possible reaction movements, when coupled to the phlogiston concept, can nicely account for the observable correlations involved in combustion and respiration. But what of the opposite reaction movement, combination?

Smelting

Smelting, that is, the production of metals from their calxes (ores), is a prime example of phlogiston combination-type reactions. Metallic calxes, such as iron or copper ores, are almost pure earths. This is obvious. The smelting reaction transforms these earthy substances into metals, mainly through the combination of phlogiston and its fiery qualities within the preexisting earthy matrix of the calx. That the transformation must involve phlogiston is evident from the fact that all smelting takes place in a fiery atmosphere, and indeed, smelting cannot occur without fire. A further point strengthens this evidence. As I mentioned earlier, metallurgical recipes always specify that the original amount of ore be intimately mixed with an addition of charcoal. There is no known metallic calx which can be smelted without addition of charcoal. (This fact was to change just as Lavoisier began his experiments.) Charcoal's rich supply of phlogiston evidently adds that extra amount of phlogiston required to carry out the job. The diagram in Fig. 5-2 depicts how smelting works.

As in the combustion reaction, smelting produces fixed air—which is merely a by-product of the main reaction process. Metals, as you can see from this explanation, are compounds of certain earths and phlogiston. Phlogiston's presence in metals is thought by many to account for the fact that the metals, unlike most other substances, have a gleaming shine. Fieriness, when captured in the earthy matrix of the metallic calx, expresses itself as the familiar shine of the metal.

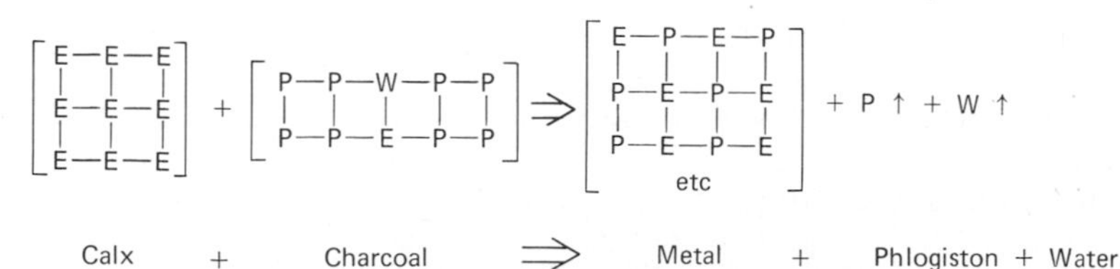

Figure 5-2. Hypothetical picture of smelting. The calx, an earthy substance (E), is combined with some of the phlogiston (P) in the phlogiston-rich atmosphere provided by the charcoal/fire environment. Water vapor (W) and phlogiston are given off as by-products.

Calcination

One additional observational reaction can be explained in terms of this hypothesis about phlogiston's role in metals. It is well known that many metals slowly calcine, that is, return to their original earthy state. Iron, for example, rusts to become a reddish, earthy powder, not unlike its original calx. Bronze, an alloy of copper, eventually produces a blue-greenish, earthy powder, similar to its original ore. These observations make clear that calcination is nothing other than the decombination, i.e., the dissociation, of the earth-phlogiston compound which is the metal itself. Calcination, to put it plainly, is a sort of slow-speed liberation of phlogiston exactly identical to combustion except in reaction speed.

This, then, is the phlogiston theory explanation for the main observational phenomena involved in combustion and metallurgy. Note that the phlogiston concept provides a thorough, systematic, and consistent logical structure for all the well-known observational correlates. The economy of thought provided by, first, the phlogiston concept, plus, second, the two reaction-movement types (combination and dissociation) is evident. Moreover, the explanation has a certain elegance and aesthetic value, mainly due to the symmetry of the explanatory logic: Phlogiston is combined in smelting, but liberated in combustion, respiration, and calcination. Thus, combustion and smelting are the mirror inverses of one another, and each is explained by only one postulated substance, phlogiston itself. Let me sum up briefly the whole of phlogiston chemistry by exhibiting its theoretical postulates and their corresponding observational correlations.

I Phlogiston: theoretical postulates

- **A** All chemical substances are composed of smaller particles (corpuscules) similar to the hard, massy atoms postulated by physics.
- **B** Phlogiston: a material corpuscule associated with fire and fiery qualities.
- **C** Dissociation: a movement of particles away from one another.
- **D** Combination: a movement of particles toward one another.
- **E** Air has a saturation point with respect to phlogiston.

II Observational correlations, plus their phlogiston explanations

- **A** *Combustion (a dissociation reaction)*

  Observational correlation:

  $$\text{Charcoal} \overset{\text{(burns)}}{\Longrightarrow} \text{earthy substance} + \text{water} + \text{fixed air}$$

  Phlogiston explanation:

  $$(\mathrm{P} + \mathrm{E} + \mathrm{W}) \Longrightarrow (\mathrm{E}) + (\mathrm{W}) + (\mathrm{P} + \text{air})$$

- **B** *Smelting (a combination reaction)*

  Observational correlation:

$$\text{Calx} + \text{charcoal} \overset{\text{(heat)}}{\Longrightarrow} \text{metal} + \text{water} + \text{fixed air}$$

Phlogiston explanation:

$$(E) + (P + E + W) \Longrightarrow (P + E) + (W) + (P + \text{air})$$

As you have probably noted, there is a certain sort of conservation involved in these reaction diagrams. That is, the qualities and various substances involved are not lost in the reactions. Rather, each is still accounted for, although in a different mix following the reaction. We might call this feature of phlogiston chemistry "conservation of qualities"; this is opposed to the "conservation of quantities" soon to be introduced by none other than Lavoisier himself.

The theory described above is essentially what Lavoisier himself learned when he studied chemistry. It is the scientific paradigm which provided the fundamental set guiding him and all his colleagues in their explorations and research into chemical phenomena. Moreover, it is this paradigm which, even as it guided Lavoisier, came paradoxically to be the paradigm he revolted against. Let me now try to resolve the paradox by showing how Lavoisier worked according to the dictates of phlogiston theory, yet at the same time produced a completely new, vastly different total view of the identical observational correlations.

LAVOISIER'S IDIOSYNCRATIC CONCEPTS

As I pointed out earlier, the paradigm which scientists use exerts a large degree of control over what they do in their scientific work. It controls particularly the kinds of questions they ask when they set up their research. Further, it sets up their expectations about what they will observe as the reaction takes place. Indeed, the paradigm ultimately specifies which results are acceptable and which are not. If unexpected results occur, their significance is evaluated almost entirely in terms of how they fit the earlier expectations generated by the paradigm. In Lavoisier's case these factors are quite evident; however, given hindsight, it is also possible to see that Lavoisier had at least three specific non-phlogiston-paradigm tendencies operating in his methods. The first of these concerned the weight balance, the second the balance sheet, and the third practical experience.

The Weight Balance

Weight balances assumed monumental scientific importance only after Newton had published his theory of motion and gravitation in 1686.[6] In

[6]Isaac Newton, *Metaphysical Principles*, F. Cajori (trans.), Vol. 1 (Berkeley: University of California Press, 1966).

Newton's theory, gravitational mass, that is, the weight effects of mass in gravitational field, had been coupled to inertial mass, that is, the tendencies of bodies to preserve their motion or rest unless acted upon by an external force. Together, these two ideas produced a synthesized concept of "mass" and its relation to the quantity called "force" ("mass $\times$ acceleration" and "$mass_1 \times mass_2/distance^2$"). Newton's third law in particular had generated the idea of a mathematical force equation. The third law stipulated that the force quantity which applied to the first body in a reaction was equal, but opposite, to the force quantity applied to the second body in the reaction. That is, each and every physical reaction involved equal, balanced, and opposite forces. When Leibniz postulated the existence of another entity, a quantity called *vis viva*, it also was believed to be conserved in force reactions. *Vis viva* later came to be understood as mv^2, or kinetic energy; eventually this entity began to figure very significantly in mathematical equations of its own. Thus, on both these fronts, mass and its interactions became the subject of close scrutiny; weighing objects and keeping close records of the mass balances and equations involved in physical reactions came to be of crucial importance.

Since some of the physicists involved in this group of mass-interested scientists were also chemists, it is not at all mysterious how some slight tradition of attention toward mass/weight became institutionalized in chemistry. However, this tradition did not come to much. Boyle had used the balance during some early researches (1670s) on calcination. But not much resulted, even though Mayow (1674) had commented on his work. It was not until Joseph Black, the young Scottish medical student, used the balance in his experiments on indigestion remedies that it came into the spotlight in England as a specifically chemical tool. But Black's work was contemporary with Lavoisier's initial research, and given Lavoisier's impending prominence, it seems fair to say that Lavoisier was the first chemist to successfully and publicly make the balance the fulcrum of chemical method and theory. How this came about is worth a closer look.

In 1764 Lavoisier submitted his very first memoir to the French Academy of Science. The memoir described his experiments with the reaction involved in the hardening of plaster of paris. Lavoisier showed via extremely close attention to the weights involved that plaster of paris hardened because of the actual chemical combination between the plaster powder and the water added to it at the time of mixing. As he noted, the plaster had originally lost a certain definite mass of water when it was prepared by heating gypsum (a naturally occurring earthy substance). When it was heated, the gypsum became plaster of paris, ready to be mixed with water in order to be molded into a hard object. The amount of water originally lost by the gypsum when it was heated was exactly identical to the amount of water mixed with the plaster. Lavoisier, in this research, had effectively used a *weight equation* to describe the reaction. This was one of the very first times that such an equation had been successfully developed to describe a chemical process. The significance of this result for both Lavoisier and modern chemistry (which comes to the same thing) is incalculable. Consider why: The phlogiston model, the last remaining vestige of Aristotelian chemical theory, did not grant any significance at all to quantitative features of chemical reactions. Rather, chemists addressed themselves all and only to qualitative matters. But Lavoisier, contrary to this strong theoretical thrust, expressed an attention to quantitative detail—especially weight/mass quantities—which is strictly inexplicable in terms of his belief in phlogiston theory. How can we account for this undercurrent of quantitative interest? We cannot settle this question in any sure detail, but several speculations reveal themselves to be very plausible.

The Balance Sheet

In the first place, Lavoisier's formal education had been in law and business. His father before him had also been a lawyer. Certainly this preparation had set him up with an expectation that scientific results would have their own "balance sheets," and moreover, that description of scientific events would be of prime significance. This tendency in Lavoisier was emphasized when he joined the French national institution of tax collecting and accounting (the Farmers General) early in his career (1768). Moreover, he became as well known, later in his career, for his success in economics and business theory as he was for his scientific results.

Second, Lavoisier's chemical education, which he did not get formally, but rather during his spare time while in college, consisted for the most part in attending the lecture-demonstrations of the famed Rouelle. Rouelle had become notorious for his iconoclastic attitude toward chemical theory. His job in the chemistry classes was to set up empirical demonstrations of the points lectured on by the teacher, who was always a chemical theoretician. Rouelle consistently demonstrated empirical effects which were opposed and, indeed, contradictory to the theoretical conceptions being espoused by the lecturer. It is not all strange that Lavoisier quickly learned the underdetermination of theory by fact. This sort of education would certainly not tend to produce a hidebound traditional mind. I also think it peculiarily important that Lavoisier was not a formal student of chemistry. All students realize the intellectual pressures put upon them by the paradigm of their study: The chemistry major, in order to graduate, must learn and publicly espouse the paradigm professed by the chemistry department's faculty. Since Lavoisier was not a chemistry "major," he did not experience these pervasive conditioning pressures while he learned chemistry. Thus, although he must need be a phlogistonian, since that was the theory of the day, he was not necessarily provided with the severe conceptual "tunnel vision" that could be expected among his contemporary chemistry "majors."

Practical Experience

Lavoisier's early scientific experience was particularly slanted toward the empirical and productive aspect of chemistry. His first real laboratory fieldwork was carried out under the auspices of Professor Guettard's geological surveys of France. On these excursions, Lavoisier was especially compelled toward careful observation, toward the quantitative accounting of mineral locations, types, and amounts. Certainly, given that geology is an intrinsically (indeed, excruciatingly) empirical science, it is impossible that this association with empirical measurement would not inculcate in Lavoisier's already predisposed mind an appreciation for the quantitative, especially in terms of masses and weights. Along these same lines it is useful to remark that Lavoisier, already by 1768—within four years of his initial plaster-of-paris balance research—had become director of the French National Arsenal and thus responsible for the national production of gunpowder. At this time Lavoisier set up his laboratory at the Arsenal, and there he had constructed what was probably the finest weight balance in the world. Production (as opposed to theory) of any chemical compound requires that the closest attention be paid to reaction amounts. Quality control demands close scrutiny of empirical detail. Lavoisier and his balance provided both. Almost immediately Lavoisier's directorship paid off in improved consistency and production of the much-needed powder (France was seemingly *always* at war during this period, usually with Britain).

As an interesting sidelight to all this, I must make some mention of Lavoisier's role in the American Revolution. American success in that war was many times due to the superiority of American arms, itself especially due to far better gunpowder—which, of course, came mainly from Lavoisier's production facilities. Even when American plants came into production, the French chemist's influence was present in the person of du Pont, a transplanted former assistant from Lavoisier's Arsenal. General Knox's well-earned reputation as an American artilleryman can be laid in no small amount to the fact that he, and not the British, had use of the best powder in the world, Lavoisier's finest! But enough of the American Revolution. Back to chemical revolution.

When we ask ourselves how it was that Lavoisier, and not any of his contemporaries, was the only one able to successfully revolt against the prevailing chemical paradigm, phlogiston theory, the features I have mentioned must be included in the answer. Lavoisier's business orientation as a predisposition toward balanced quantitative representations, his iconoclastic and informal education at the hands of Rouelle, and his thorough grounding in the empirical necessities of weight measurement together provided sufficient conceptual impetus to propel his eventual drive against phlogiston theory. Let us now take a careful look at the actual details of his first moments of revolt against the accepted theoretical establishment of chemistry.

EXPERIMENTAL INSIGHT

In the early 1770s Lavoisier, along with most of his contemporaries, especially Joseph Priestley in England, began to focus his attention particularly closely on the group of reactions central to the phlogiston theory. They all became extremely concerned with the phenomena surrounding combustion and smelting. Lavoisier's most crucial research began in 1772. It focused upon the combustions of sulfur and the newly discovered substance phosphorus. His first move, as my earlier comments would lead you to expect, was to weigh the materials before and after burning. The results violated his expectations, and quite surprised him in a crucial way. "About eight days ago," he wrote, "I discovered that sulfur in burning, *far from losing weight, on the contrary*, gains it; it is the same with phosphorus."[7] Lavoisier's surprise here is quite evident. His expectations clearly were that sulfur and phosphorus would lose weight upon combustion; but the results were far from that, "on the contrary" weight was gained. Consideration of the tenets of phlogiston theory will quite explain Lavoisier's surprise. To express the situation symbolically:

C__: The substance ____ is undergoing combustion.
G__: The substance ____ gains weight during combustion.
p: The substance phosphorus.
s: The substance sulfur.

These symbols can now be used as follows to express the correlations which Lavoisier observed in his experiment. First, we see that

"Cp.Gp" is a correlation (1)

and second, we see that

"Cs.Gs" is a correlation. (2)

[7]Conant, op. cit., p. 16.

Lavoisier, making use of these correlations, and others that he knows, reaches a kind of intermediate view which may be expressed as an 'if, then' conditional:

$$(x)\ (Cx \rightarrow Gx) \tag{3}$$

which, in plain English, is the claim that all substances, if they are burned, gain in weight during combustion.[8] With this well in mind, it is now possible to account for Lavoisier's surprise that sulfur gains weight during combustion. Think back to what was said earlier about the explanation of combustion according to phlogiston theory: Combustion is a *dissociation* reaction, a process in which the burning substance *loses* phlogiston. We use the following as symbols:

C___: The substance ___ is undergoing combustion.
L___: The substance ___ is losing phlogiston.

We may now represent the explanation of combustion according to phlogiston theory as the universal law:

$$(x)\ (Cx \rightarrow Lx) \tag{4}$$

This law states that all substances which burn lose phlogiston in so doing. We can now combine symbolic statements (3) and (4), and using some straightforward deductive logic—as Lavoisier also did—come to the same surprising conclusion he did. Expressed in symbols, a result of combining (3) and (4) is that

$$(x)\ [Cx \rightarrow (Gx \rightarrow Lx)] \tag{5}$$

This states that, for all substances, if they are combusted, then, if they gain weight, they lose phlogiston. Weight gain during combustion is the result, according to this theoretical causal account, of the loss of phlogiston. In plain speech, loss of mass produces gain in weight. Lavoisier certainly deserved to be surprised by this result. He, a most empirical man, a man led to expect balanced numbers by his background in business, physics, geology, and practical production, is equally led to conclude, "Mass loss produces weight gain."

Lavoisier, at this point, is faced with a monumental choice, indeed a choice that has incredible significance for the history of chemistry. His

[8]Lavoisier, in print, seems to restrict this claim to the metals. But he obviously does not strictly *mean* the restriction, since sulfur is not a metal.

paradigm, the phlogiston theory, can be used to explain what is going on in those combustion reactions, namely, the assertion "Combustion produces weight gain via mass loss." This proposition is completely consistent with the theory. In this way, Lavoisier's precise new facts—the weight gain of combusted sulfur and phosphorus—can be "accreted," tied into, the old theory. However, the new proposition is not without its own implications, one of which Lavoisier finds very difficult to accept. You can easily see the troublesome idea: If loss of a substance causes weight gain, then the lost substance must have a negative weight. Only if the burning substance loses a negative quantity can it gain a positive quantity. On this account, phlogiston must have a negative weight.

The Negative-Weight Problem

To the modern mind, the notion of a negative weight sounds implausible, even silly. However, this is not necessarily the case, especially when we consider the situation from the point of view of Lavoisier and his contemporaries. Try to see it their way: The phlogiston theory is well verified, and it has quite a bit of explanatory power in that it can explain many of the observable correlations involved in smelting, combustion, and so on. Thus the logical and psychological pressure is on Lavoisier to accept the proposition "Combustion produces weight gain via mass loss" and its implication "Phlogiston has a negative weight," simply because these propositions can be snugly fit into the tradition of the phlogiston paradigm. Moreover, there are certain benefits which can be realized by agreeing to accept "Phlogiston has a negative weight." Let me describe just one for you.

At this very same time in France, the Montgolfier brothers were beginning their hot-air balloon ascents over Paris. Phlogiston theory could have been used to explain the mechanics of their flights. Consider the following. Phlogiston has a negative weight, and thus its force is directed away from the earth's gravitational center; hence, phlogiston will fall *up* in the terrestrial vicinity, rather than down. If phlogiston could be trapped and concentrated together, then perhaps this lifting force could be used to do work. Suppose, for example, that someone were to manufacture an enormous bag, and to kindle a very hot fire beneath this bag in such a way that the bag trapped the phlogiston, for however briefly, as it flew away from the fire. Is it not obvious that, if enough phlogiston could be trapped, the bag itself would soon float away? Then, if one (or two) were brave enough, one could suspend a passenger basket below the bag and *Voila*! fly away. Because of phlogiston's negative weight, we could perhaps call this vehicle a "lighter-than-air" balloon.

However, in addition to this further extension of the theory's explanatory power, the new principle "Phlogiston has a negative weight" has two further things going for it, one historical, one metaphysical. In the first place, historically, phlogiston's conceptual roots are forged in fire. And fire, as one of its elemental behaviors, naturally goes up. (Have you ever seen a fire burn *down*? That is, without the help of wind or something else?) Given phlogiston's relation to fire, one can only expect that its force vector points up. Second, consider the causal problem. As we all well know, it is difficult under any circumstances to accept a causal hypothesis without at least some dim clue about the actual mechanics of the causal relations which are involved. The mechanism postulated by this hypothesis must always be at least imaginable before the explanatory principle using it can be accounted "acceptable." Thus, if someone said to you, "Event E occurred because it was caused by process C, although there is absolutely no conceivable mechanism in nature which could be responsible for C," you would be hard pressed to give even initial consideration to the "E because C" hypothesis. Thus, in order to give even initial consideration to the proposition "Phlogiston has a negative weight," there must be some conceivable mechanism which could account for the property "negative weight." As it turns out, there are at least two which can be given in the terms of Lavoisier's time.

First, there is one related to what I just said about phlogiston's conceptual roots: If fire was once thought to have natural "levity" (lightness), then levity might be a natural property of phlogiston. On this account, levity is just as natural as "heaviness." After all, natural philosophy has led us to expect that nature has certain symmetries: hot and cold, plus and minus, matter and antimatter, and so on. Why not a heaviness/lightness symmetry as well? Surely there is no general philosophical reason against it. Thus, in order to provide a mechanism for phlogiston's upward movement—its negative weight—we can simply postulate a natural property called levity. Some of Lavoisier's colleagues did exactly this.

But a second alternative also exists: the principle of buoyancy. Archimedes' discovery of the mechanism whereby less dense bodies float in denser media had been rediscovered and was well known by Lavoisier's time. Certainly it offered itself as a plausible candidate to provide levitating phlogiston's mechanical support. Although the details would need some working out (for example, was levitating force exerted by phlogiston while it was combined with the calx in metals?), they need not appear stultifying to the phlogiston theory. Buoyancy offered a plausible mechanism for phlogiston's negative weight.[9] For reasons like these,

[9]To see an example of this, see the article "Guyton de Morveau Argues the Relative Levity of Phlogiston (1772)," in M. P. Crossland, *The Science of Matter* (Baltimore: Penguin Books, 1971), p. 735. De Morveau was a well-known scientist of the times. His defense of phlogiston is both significant and fascinating.

many of Lavoisier's contemporaries accepted phlogiston theory as modified by Lavoisier's new and accurate results. In their minds, the phlogiston tradition had moved ahead—albeit a bit oddly—through Lavoisier's work. This, after all, is the way that paradigms grow into vastly general traditions: They pick up new facts and either quickly snuggle them into the overarching network of already-accepted explanatory concepts, or, if necessary, do a bit of cutting, filing, and shaping up of the parts of the whole system where required to fit in the new items. In these processes the paradigm grows ever bigger and more powerful.

Lavoisier, however, did not mesh "Phlogiston has a negative weight" into the paradigm he was carrying around in his head. Rather, when he got to the antecedent proposition (5) ("If combustion and gain in weight occur, then phlogiston is lost"), he stopped cold in his tracks. And then, apparently without even a pause for a quick breath, he rejected the entire phlogiston theory, part and parcel. We see his own words:

> This increase of weight arises from a prodigious quantity of air that is fixed during combustion and combines with the vapors. This discovery, which I have established by experiments, that I regard as decisive, has led me to think that what is observed in the combustion of sulfur and phosphorus may well take place in the case of all substances that gain in weight by combustion and calcination; and I am persuaded that the increase in weight of metallic calxes is due to the same cause.[10]

Here in this very quote Lavoisier throws out phlogiston theory, and announces at the same time a main idea of the replacement he will eventually offer for the venerable old conceptual system. But since he does not explicitly assert that he is doing so (he certainly does not say "And I hereby dub phlogiston theory 'dead'!"), let me make his moves a bit more explicit.

Complete Reversal of Concepts

Note first that he puts the weight gain down to a "combination." That is, the burning substance gains weight because it *combines* with another substance. That seems clear enough. However, now remember the phlogiston explanation of combustion (and calcination as well): Combustion is a *decombination* reaction in which the burning substance *loses* phlogiston. According to phlogiston theory, combustion is a *dissociation* reaction. But Lavoisier here denies this; in contradiction, he asserts that combustion is a *combination* reaction. Thus, his new explanation is the exact opposite of the prevailing paradigm; and in offering it he expressly opposes the tradition. Moreover, his opposition at this one point, his

[10]Conant, op. cit., p. 17.

offering this single, unique, contrary proposition, spells the logical doom for the phlogiston paradigm. To see this, recall my earlier depiction of a paradigm: It is a species of conceptual system. Conceptual systems are *systems* just insofar as there are logical relations between the concepts involved. Together, the totality of concepts in a conceptual system are woven into a solid, interrelated logical *network*. If you too much modify any single proposition, the whole interwoven network is put into jeopardy. In the present case, Lavoisier does not just *modify* a concept, he contradicts it, he destroys it, and he replaces it with its mirror opposite: Combustion is not dissociation, it rather is combination. Such a drastic move will have a dear price, namely, the whole of phlogiston theory. Every principle, every dovetailed element of the system is by necessity changed. Take one clear example: The first concept to expire is that one which is symmetrical to "Combustion = a dissociation reaction," that is, "Smelting = a combination reaction." Smelting must now be totally reconceived—reversed, in fact—to become viewed as a dissociation reaction. Further implications are even more staggering. Formerly, metals were thought to be nonsimple substances since they were compounds of calxes plus phlogiston. Calxes, on the other hand, were simple elementary substances—specific sorts of earths. All this has now been reversed, if Lavoisier's idea is correct. It is plain to see that phlogiston theory cannot survive these changes in any recognizable form, given acceptance of Lavoisier's notion that "Combustion is a combination reaction."

One further epistemological consequence of Lavoisier's move must be noted. As I explained earlier, problems and methodologies are controlled by paradigms. Thus, if scientists believe paradigm *X*, they will concentrate their research solely on questions which make sense in terms of *X*. Their methodology will be adjusted to produce the types of data which are logically consistent with *X*, and so on. This natural implication of the logical structure of a paradigm does much to explain Lavoisier's new research directions in the period immediately following that described above. If Lavoisier's new concept group (we can hardly call it a concept structure, let alone a paradigm, at this early stage of the game) contains the notion that "Combustion = a combination reaction," then a necessary logical consequence would be "There is a 'something' which combines with substances when they burn." Herein is a new problem posed specifically by the new set of concepts. According to the phlogiston theory—since it does not assert that "There is a combining 'something' involved in combustion"—there is no reason to search for that "something," no reason to try to identify it and classify it among all other known substances. But for Lavoisier, it becomes extremely important that he identify what it is that he believes combines with various substances as they burn. In fact this new research direction consumes Lavoisier for the next five years, as we shall see in Chapter 9.

From Quality to Quantity

However, a further comment is in order about the new directions Lavoisier turned chemistry toward. Phlogiston chemistry, from the very start, is a qualitative explanatory system. I have already suggested some psychological and historical reasons why Lavoisier even in the beginning was somewhat antiparadigm insofar as he showed interest in quantitites. His new concept set only reemphasizes and more tightly focuses this original concern. As we shall see, Lavoisier continues to use his balance and his balance sheet, and in fact soon produces the very first chemical equation. Chemistry thus gains a quantitative power because of Lavoisier's new ideas. But his move does not produce only profits. Something was lost from chemistry as well. In the former tradition, attention had been directed toward the qualitative. Explanations of qualitative properties were sought; for example, metallic shine was linked to the fiery qualities of phlogiston, and the earthy features of calxes were accounted for by their earthlike composition, and so forth. All this went by the board when Lavoisier ushered in the new numerical age of chemistry. For this reason, we must always be careful not to simply equate paradigm change with progressive increase in explanatory scope. Paradigm change always involves *some* perceived improvement (otherwise, why change?), but there are always some losses as well. When paradigms change, so do the implied questions, problems, and solutions which occupy scientists. Usually involved in the situation is some question, problem, or solution whose loss represents a sacrifice, at least in the intellectual component of the science. But notwithstanding the loss, "science marches on" in some important sense during paradigm change. It is no different in this present case, as we shall see. But first, there is one glaring difficulty left to be confronted.

THE UNITY OF SCIENCE

Lavoisier, unlike his contemporary chemists, came to a complete and screeching halt in the face of "Phlogiston has a negative weight." The essence of his discovery is contained in this halt. The obvious question which remains is: Why did he alone behave like this? The answer—as in all interesting cases—is not a simple one. As I noted earlier, Lavoisier's education was a peculiar one. He was not a "professional" scientist, one trained formally and explicitly in the theories and techniques of a scientific discipline. This left him somewhat "looser" than his contemporaries in terms of intellectual constraint. Second, his "amateur" standing had been gained in the best educational circumstances: His classroom and field experience environments were peopled by the best scientists of France. Further add to this his business and legal preparation. Together, these factors provide the impetus which propels him into use of the balance, with an attendant care for the quantitative. But it seems clear to me that one further overriding intellectual feature is needed to provide sufficient explanation of his revolutionary move against the phlogiston theory. This feature, when coupled to those previously mentioned, provides a clear explanation of his revolt.

Lavoisier's overall education provided strict training in the best of classical French style, a style which includes long and intense steeping in "letters"—particularly those of Cartesian rationalism. This style, although it is often gazed upon nowadays with a dour expression, involves the highest commitment to what the French call *la logique*—a special predilection and motivation toward the systematically abstract and theoretical. Descartes, the first (and in some sense the "ultimate") rationalist, provided the exemplar for French thought-style. Descartes calls for the deductive systematization and unification of all knowledge, and calls specifically for the unification and consolidation of science. An abstract logical structure is ideally to be forged, a single supporting structure upon which to array and display all scientific thought. It is simply inconceivable that Lavoisier could have escaped both the penchant for the abstract—*la logique*—and the expectation of the eventual intellectual unity of all the sciences.

Given this sort of intellectual molding in his education, Lavoisier acquired his full knowledge of the Newtonian physics, especially its reliance upon the notion of a gravitational mass. The very idea of a physical object includes its gravitational mass in the Newtonian theory. Only via this idea can one apply the force equations which are the foundations of dynamics. When Lavoisier began his chemical studies—and it is essential to recognize that *chemistry at this time was a tradition separate from that of physics*—his expectations must have been that somehow, in some fashion, sometime, all the sciences, including both physics and chemistry, would be one. But he could not keep *this* mental set, this particular epistemological principle about the unity of science, in the face of, on the one hand, a physics which asserted "All objects have gravitational mass," even while, on the other hand, chemistry asserted "Phlogiston has a negative weight." Something had to go, and clearly it was not going to be Newtonian physics, a science which could boast of nearly a century of spectacular success. Thus, traditional chemical theory, that phlogiston paradigm which could include the concept "Phlogiston has a negative weight," had to fall in order that it might be replaced by a new paradigm, a paradigm more congenial to Newtonian physics.

It seems to me that we must postulate Lavoisier having a belief such as this Cartesian one if we are to explain the chilling ease with which he disposed of seventy years of chemical thought. "*Something* must have combined with the burning substance in order to cause its weight gain." Lavoisier dropped this idea on himself, on his contemporary chemical community, and finally, on us, with inexplicable ease. The move will not, simply cannot, be explained in terms of some other of his beliefs; and this Newtonian notion that "Physical objects have gravitational mass" is supremely qualified to fit our explanatory need. To us, in a modern tradition permeated by 300 years of Newtonian commitment to mass and weight, Lavoisier's belief that it *must* be a combination reaction seems to be the only reasonable move. But this apparent reasonableness is due solely to our *own* beliefs. Once we try, really try, to put ourselves in Lavoisier's head, the reasonableness of the move must still be there, but *not* in terms either of our 300 years of Newtonian masses, or of the chemistry he and his colleagues practiced. Rather, it must be reasonable in terms of some cluster of particular beliefs which Lavoisier held to, a main one of which is the Newtonian view that "Physical objects have gravitational mass."

Some Objections

Of course objections can be raised against my argument here. For example, a main one would be straightforwardly historical: We do not know for sure that it was his commitment to Newtonian physics and his belief in the Cartesian ideal which ultimately propelled Lavoisier to reject the phlogiston model. There is a certain amount of force to this objection; however, I believe that we ultimately must reject it, and for several reasons. In the first place, it is doubtful whether or not we can *ever* be rock-solid certain about all the necessary conditions which caused a creative act in a scientist's mind (or anyone else's mind, for that matter). Thus, the best that we can do is to provide a probable cause, a cause which is sufficient to explain the actual event. This brings us to the second point: My explanation, if true, is certainly sufficient to explain Lavoisier's rejection of the phlogiston theory. Obviously this counts in its favor. Third, given what we do know for certain about Lavoisier's education, my explanation is not only sufficient, it is genuinely plausible. Not only could the discovery have happened the way I describe it, but also it is likely that it happened that way. For these reasons, I think that I am justified in recommending my explanation as an adequate one.

There is an interesting methodological point brought up in all this.

Whenever we try to analyze the activities of historical persons—especially scientists—we are faced with the problem of discovering what *really* went on. In terms of accounting for creative acts this is exceedingly and peculiarly difficult, since what we are trying to do is to perform a sort of "long-distance" logico-psychological investigation into the goings-on of a human mind—a difficult procedure even at firsthand. In terms of our own interest, however, there are certain things which ease our task because they do not seem fair game for our investigation. For example, our interest is sharply different from that of the psychoanalyst. He probably would ask questions about Lavoisier's sibling rivalry, about how his father treated his mother, about his early eating and other habits, and so on. But these things are not really to the point for the student of science. Rather, it seems to me that we are after the *rational* elements which influenced Lavoisier. Thus, if I may be allowed to distinguish between conditioning and learning, between training and education, I would say that study must focus upon the latter member of each of these pairs, upon the sorts of things discussed above: commitment to certain logical and metaphysical goals, certain sorts of learned expectations, and so on. Luckily for us, these things are much more discoverable than the kinds of activities and events that the psychoanalyst must seek. But our task is not without its own peculiar difficulties. Take only one example. It would be really a clincher if we could find written down somewhere in Lavoisier's own hand, "My commitment to Newtonian physics and the Cartesian ideal led me to reject phlogiston theory." But nowhere can we expect to find this sort of statement, unless it be in something like his autobiography or memoirs. Unfortunately, Lavoisier's life was chopped off before these books could be written. But even if they had been written, there would still be a problem. Memoirs are usually written toward the ends of careers, and most especially they are written only *after* successes have elevated the memoir writer to the exalted state where he can expect that someone else might even care to read the work. Problems with this timing are obvious: Time dulls memory, self-reflection over the years often leads to reporting the creative acts as more structured than they probably were, and so on. For these sorts of reasons, memoirs must be taken with a grain of salt.

Another source of data is the "official" report of the outcome of the creative act. This is the sort of information provided by the passages quoted above. But there are problems with this kind of thing as well. The "official" report is just that, it is official. Thus, it is more precise, more formal, and definitely more clearly worked out than was the original discovery. Although it is nearer to the discovery in time than is the memoir (and probably much less reflective and "philosophical"), it must necessarily be couched in terms and a style acceptable to the contemporary scientific community. In order to get around these sorts of difficulties, one must do just what I have done above, namely, look at the report, and first, try to fill in the gaps in reasoning. Thus, I tried to explain what Lavoisier meant by his use of the term "far from" in the phrase "far from losing weight." But this is not enough; one must also attempt to fill in as much as possible of the rational background to the passage, since it cannot be really meaningful until it is portrayed within its full context. Hence my description of Lavoisier's educational and philosophical commitments. Once the whole of this is done, we are faced with a probable and plausible explanation as a product. This explanation is the best that can be made. But it seems to me to be a good one, since it satisfies our main desire, which is to understand and describe the process of scientific discovery. I am not claiming that, simply because we understand some of the elements surrounding Lavoisier's particular discovery that phlogiston theory needed rejection, we thus understand what is involved in scientific discovery in general. There are many more elements remaining to sort out in the general case. However, one thing which should be clearer is the plausibility of my initial claim about discovery: Discovery in science occurs within a paradigm, but it occurs because of conflicts between the ideas generated by that paradigm and the ideas generated by other, nonparadigm conceptual structures. I will argue this general point in two more cases, that of Pasteur and that of Pauli, in an effort to see just how far it can be pushed.[11]

The Essence of the Discovery

But now back to Lavoisier. The crucial move occurred in 1772, in the work described by the quoted material above. Lavoisier himself later admitted that this was the crucial moment, and so it seems to me that we can all agree that it constituted what can be called "Lavoisier's discovery." But the obvious question remains: *What*, indeed, had been discovered? This is not a simple question to answer. However, it appears that at least two things were involved. The most significant one is a logical point: Phlogiston theory had to be rejected because it could no longer be considered consistent with other conceptual systems believed by Lavoisier. The second point is a more straightforwardly scientific one: The processes involved in combustion are combination processes, and not dissociation processes. This latter point provides some content to Lavoisier's rejection of phlogiston theory. That is, he now had a rough idea what to look for in order to come up with a replacement for the phlogiston concept; he must look for the "something" which does the combining

[11] I will modify it slightly in the Pauli case, in order to accommodate cases of conflict between *levels* (e.g., more versus less fundamental concepts) of a paradigm.

during combustion. As always, the idea—even the rough concept found in the first rough stages of the new paradigm—leads the expectations, the mind sets up new programs of research for the hand and eye.

A more traditional account of the whole Lavoisier case would be different from mine. It would focus upon the period 1775–1776, when the substance variously called "de-phlogisticated air" or "oxygen" or even "the healthiest and purest part of air" was actually verified to exist by Priestley, Lavoisier, and Schele. But this does not seem to me to be the *real* discovery that Lavoisier contributed to this epoch of chemical history. The *real* discovery occurred when he postulated the existence of a substance such as was later found. That is, a real discovery is contained in the postulate of its existence, not in its actual verification. To a great extent I want to say that the substance was found and recognized only because it was postulated. The view I express here is a somewhat tricky one, since it strictly implies that accidental discoveries cannot be made in science. Many would disagree with my claim. But I think that it is defensible; and perhaps this is the best place to defend it.

ACCIDENTAL DISCOVERIES

Accidental discovery is the sort of thing which many think happened to Columbus when he stumbled across America as he was looking for India. But this is not a good model for scientific discovery. Think back to what I said earlier about the "mental set" provided by paradigms. There I claimed that Plato's paradox of discovery could only be answered by recognizing that the conceptual systems already in our heads prepared us to recognize and perceive the objects presented to us by the world. Thus, no paradigm, no foreknowledge; no foreknowledge, no preparation; no preparation, no recognition; finally, no recognition, no perception. And to say that someone has "discovered" something seems to presuppose that the someone can recognize and perceive what the "something" is. This is true even in Columbus's case since he certainly had some clear idea about what a "non-European continent" would be like if he ever came across one. But the scientific case is even clearer.

A good example is provided by what many take to be the ultimate "accidental" discovery in recent science, Charles Goodyear's discovery of the vulcanization of rubber.[12] The way the story is told, Goodyear by accident spilled some sulfur into a vat of gooey India rubber (a substance just like art gum erasers) he had heating on the stove. When the sulfur combined with the hot sticky mass it hardened into the substance we now

[12]Ralph F. Wolf, *India Rubber Man: The Story of Charles Goodyear* (Caldwell, Id.: Paxton Printers, 1939).

know as rubber. Eureka! Goodyear had accidentally discovered how to harden India rubber, and make it for the first time commercially useful. But this story is farfetched. In the first place, why in the world would Goodyear have had the rubber heating on the stove? Why would he have some sulfur about? How did he know that it was valuable to have hard rubber rather than gooey rubber? And so on, and so on. The truth is, Goodyear had been looking for a means to harden India rubber for over three years. He had expended his entire fortune plus that of his family and friends in his search for some way to make India rubber hard enough to use in commercial products. He had from the start believed that the theory of alloying metals (according to which certain substances are combined with the metals in order to produce various desired properties) was relevant to figuring out how to harden rubber. Thus he had sought for some substance which would be a suitable "alloy" for the sticky rubber he was working with. It was no accident at all that Goodyear discovered the process we call *vulcanization.* He was looking for such a process, he knew quite well what it would be like to succeed at his task, and he even had a good clear paradigm—the conceptual system of metal alloying—as a guide for his research.

An entertaining example may serve to bring this point finally home. Suppose that, rather than Goodyear, it had been a janitor doing some cleaning (let us assume that Goodyear has not so impoverished himself that he cannot afford to have his laboratory cleaned) who spilled the sulfur into the rubber. Thus, while he was doing the nightly mopping, his mop tipped the sulfur tin and its contents poured into the simmering gooey mess of whatever it was Mr. Goodyear was fooling around with. The gooey mess immediately hardened into a stiff blob. What to do now? He has obviously ruined whatever it was that Mr. Goodyear was doing, and the best next move will be to cover up. So he takes the hardened blob and throws it away somewhere where it will never be found. What now about the "accidental" discovery of vulcanization? Can we say that he has accidentally discovered a way to harden rubber? It does not seem to me at all that we can say this. What he has discovered is that he had better be a darn sight more careful when cleaning up around Goodyear's lab, unless he wants to lose his job.

My point here is not a complicated one. I want to claim that discoveries can happen only to those whose conceptual systems have somehow prepared them to recognize what it is that will be eventually discovered. And even more strongly, I claim that in the Lavoisier case the essential discovery is the two-pronged one mentioned above: first, that phlogiston theory must be rejected; second, that the "something" which combines during combustion must be found.

An implication of this claim is that I must construe the ensuing stages of Lavoisier's work (and that of his colleagues) to be something other than "discovery." I do. It seems most plausible to describe Lavoisier's later work as *verification*, that is, the process of tying down his new concepts to empirical results in the laboratory. In this process Lavoisier had help from one other man, Joseph Priestley, the Englishman I mentioned earlier. The cooperation—and the controversy—between these two men will illustrate for you some of the best aspects of the communal nature of science. But let me save that part of the story for later. Right now, let us turn our gaze forward in time almost a hundred years. There, in the 1850s France of Louis Pasteur, we will find exhibited another interesting case of scientific discovery.

SUGGESTIONS FOR FURTHER READING

In the Suggestions for Further Reading in Chapter 4 I recommended several philosophical accounts of scientific discovery. I hereby rerecommend them to you. Especially if my recommendation did not stick the first time! In regard to some details of Lavoisier's life and studies, the best and most readable account is still Douglas McKie, *Antoine Lavoisier* (New York: Collier Books, 1962). This is an especially good book in that it provides very valuable information not only about Lavoisier's life, but also about his scientific background and theories. My account of Lavoisier's discovery is not the only one around. It has been discussed several times—in more detail, and from a different viewpoint as well. A good example of this is Henry Guerlac's *Antoine-Laurent Lavoisier, Chemist and Revolutionary* (New York: Scribner, 1975). For a solid critique of Guerlac's theory about Lavoisier, see Maurice Finocchiaro, *History of Science as Explanation* (Detroit: Wayne State University Press, 1973).

Chapter 6

Pasteur and Microorganisms

INTRODUCTION

Louis Pasteur is justly accorded the title "hero of science." However, it is not always entirely clear what people mean when they refer to him this way. Often it seems that they think Pasteur was a physician who invented new medicines to cure diseases. In a way this is true—but it is not the most essential point about Pasteur. In reality, Pasteur's real scientific heroism derives not from his medical research alone, but more importantly from his theoretical work, work which exposed him to severe criticism from the paradigm community of which he was a member. It was not until quite a bit later in his career that Pasteur came to invent the various treatments and vaccines which carry his fame today. But long before that he had become interested in a part of human life—alcoholic beverages—which was to involve him deeply in the investigations that propelled him to the forefront of his scientific community. Let me say a few words about this fundamentally human endeavor in order to introduce Pasteur to you.

The core of the matter is that Pasteur was employed by the big commercial brewers and vintners of France to find out why their bottled goods sometimes spoiled.[1] It was the investigation of this problem which eventually provided the key concept in his new paradigm, the microorganism theory of fermentation.[2] Moreover, it turned out to be only a short step from this theory of fermentation to an analogous theory of disease. I will explain later why these two apparently unrelated fields—fermentation and disease—got linked together via Pasteur's work. Beer and wine making are among the major industries in France. During Pasteur's time, approximately 30 percent of the French work force was engaged in the growing, production, distribution, and marketing of these beverages. Consequently, it is easy to see how the problem of beverage spoilage could be an important one—important enough in fact that a man of Pasteur's stature could be hired to attempt to solve it. To look ahead a bit I can say that Pasteur solved the problem in a short period of time, and his method of solution is still with us today: We all drink pasteurized milk as a direct result of his researches.

Pasteur's story is a fascinating one, and it is going to take a bit of telling in order to get all the details rounded up. I want to make clear at the outset what I believe to be the discovery for which Pasteur should actually receive eternal credit: Pasteur discovered, in direct opposition to the prevailing conceptual structure of the time, that microorganisms were responsible for fermentation reactions of all sorts. The result of this was the creation of an independent set of explanatory concepts—the theory we call *biology*.

In telling Pasteur's story, I will first relate a brief bit about Pasteur's life, just to get you familiar with him. I will outline the paradigm according to which he practiced. Then, I am going to sketch his first research project, which was involved with the analysis of molecular structure using polarized lenses (just like the ones found in sunglasses). Next I must relate some of the information involved in beer and wine making, since I will argue that it was his work on fermentation which provided the nonparadigm conceptual element in Pasteur's discovery.

As I argued in the Lavoisier case, revolution in a science comes from the collision in some scientist's mind between the going theory and some already known, but nonparadigm conceptual system. It will be quite clear in Pasteur's case just what this outside material consists of. By the time I have finished describing this material, the story of Pasteur will be about told, at least so far as the discovery phase of his work is concerned. In

[1]In point of fact, some of his main work involved the fermentation of *beet juice*. But this historical point is insignificant in terms of the concepts involved.

[2]I have related my discussion to another pamphlet from the Harvard Case Studies in Experimental Science Series. This one is Case 6, James Bryant Conant (ed.), *Pasteur's Study of Fermentation* (Cambridge: Harvard University Press, 1952). It is really a valuable little book, and I strongly encourage you to buy it (it is inexpensive) and follow along with the case.

Chapter 10 we will look at the controversial stage of Pasteur's work, that period when his new paradigm was being rapidly generalized to cover an ever-increasing range of previously disparate fields, including surgery, medicine, immunization, and agriculture. Although it is this latter phase of his work for which he usually gets the most credit, it seems quite clear that the initial discovery which we shall be concerned with in this chapter is the actual creative act which kindles the later blazing successes of Pasteur and late-nineteenth-century medicine. Let us now take a quick look at Pasteur's biographical facts.

BIOGRAPHICAL DATA

Pasteur grew up in the northeast region of France, a region called the Jura, which abuts upon Switzerland and Germany at their western borders. He was born in 1822 in the town of Dôle. When his education was just begun, the family moved to the little town of Arbois—which still maintains his family home as a monument. At the age of twenty he left for Paris to attend the Ecole Normale Supérieure. This college had been especially established to train university professors, and in Pasteur's day it provided one of the best educations in Europe. (It still does today, although now the Ecole Polytechnique probably provides a better research/technical schooling. But this was not necessarily so in Pasteur's time.) Pasteur studied physics and chemistry at the Ecole, was graduated in 1847, and within a year had a professorship in physics at the University of Dijon. Dijon, the famous mustard town, did not keep him long; apparently he desired to move somewhat nearer home. Thus he accepted a professorship in chemistry at Strasbourg, in the French Rhine River area. But soon he was on the move again, going to Lille up near the Belgian border in 1854. But this time he was not merely a professor; he was appointed dean of the faculty of science, quite a promotion. By 1857 his stature had grown to such a point that he became director of scientific studies at his alma mater, the Ecole Normale. Also at this time he was hired to do the wine and beer research. In 1863 he became a faculty member at the Ecole des Beaux-Arts, where he stayed for four years. Finally, he reached the pinnacle of French professorial success: In 1867 he became professor of chemistry at the Sorbonne, the renowned University of Paris. He stayed there for twenty-two years, and did not leave until he became director of his own establishment, the Pasteur Institute. This institute was founded in 1888 as a memorial to his accomplishments. He died in 1895 while still its first chief.

There are two remarkable things about Pasteur's career which must be noted. First, its successes. Pasteur moved upward on the academic ladder with blinding speed. This is a certain indication of his talent.

Second, we must remark that his positions were always in the physical sciences, specifically physics and chemistry. This feature points up something which is not often recognized about the departments of nineteenth-century science: Biology, as a separate science, was not formally organized until very late in the century. Indeed, as I will try to show, it was Pasteur's work which constituted a significant focal point for the development of modern biology. He himself realized this: By the time he had made his discovery, it was extremely clear in his own mind that he was establishing a science of living systems. His interest, methodology, and theorizing were concerned solely with the living entity and its activities, going against the mainstream of the chemical paradigm. In this regard we must admit that Pasteur was most certainly setting up a new science, and this in opposition to the prevailing paradigm, which regarded life as a process which fitted within chemistry. His main battle in the verification stage of his theory, as we shall later see, was with the "chemists," those researchers who were interested in life, yes, but who saw it as just one more chemical phenomenon. But Pasteur was not fighting the chemists alone. He was also fighting another tradition, medicine. Chemistry and medicine had had quite a connection in earlier centuries. Indeed, up until the time of Lavoisier, chemistry had been the handmaiden of medicine, expending much of its conceptual and theoretical effort on developing links between chemical information and human disease. Lavoisier's work was for the most part responsible for carving out a niche which was strictly chemical in its own rights. Clearly the industrial applications of precise chemical knowledge—the first real instance of this being Lavoisier's powder manufactory—had solidified this independent role. But the chemistry/medicine connection still remained in Pasteur's time even though of much less strength than in the times prior to Lavoisier. Pasteur had this link to sever, and of course, we realize now that he was completely successful in this project.

In an effort to avoid the "can't see the forest for the trees" syndrome, before getting deeply involved in the details of the case I want to broadly contrast the events of Pasteur's revolution with those of Lavoisier's revolution, specifically in regard to the question of the unity of science. As I argued earlier, Lavoisier's main move was to "physicalize" chemistry; that is, he united it with the physics of his time in terms of the methodology and epistemology of balance scales and quantitative interest. Moreover, of course, he rebelled against the metaphysically radical notion that there was a substance with a negative mass. But the upshot of all this was to connect chemistry and physics together via one fundamental philosophy. Pasteur's move, as I have already hinted, was opposite to this; he worked to carve out a new science from within the territory of an older, well-established one. Thus we can see his program as involving the "declaration of independence" of biology from chemistry. Contrasts between these two revolutions, as I will later emphasize, are very revealing of the reasons why revolution attempts either succeed or not.

THE CHEMICAL PARADIGM

Pasteur's chemical studies involved some areas of what we would now call—mostly because of the success of Pasteur himself—biology and biochemistry. But in his time, these areas were not thought to be in any essential way different from more typical and routine chemical matters. Chemical theory in the period following Lavoisier had become more and more physicalized. It was postulated that chemical substances behaved solely according to the processes of their underlying material structure. Chemical atoms had been postulated and fairly well verified. Molecules—superstructures of atoms—had been invented as explanatory concepts for many observable reactions. Finally, chemical activity had become "mechanized," that is, it had come to be construed strictly as the result of the mechanical motions of the underlying atoms. Thus the notions of combination and dissociation were fundamental in the conceptual analysis of chemical activity. During this period, life and life processes were not thought to be anything other than reducible to the physico-chemico-mechanisms recognized in scientific theory. Life processes were not conceived to be at all logically independent in their conceptualization from chemical ones. Thus, as I noted earlier, there were no professional "biologists" as we now know them. (I am here excepting the quasi-scientists of the German *Naturphilosophe* movement.) A quick look at some of the explanatory concepts of the time will reveal all we need to know about the paradigm believed by Pasteur and his colleagues.

Fermentation reactions and diseases were prime cases of hypothesized chemical effects. In the former case, some mechanical agent was believed to be present in the yeast which, when added to an unstable chemical formation such as that found in malt extract, would cause the unstable medium to go through a vigorous collapse and alteration. Diseases were thought to be somehow analogous to this. Probably the foremost theory was that propounded by the great German chemist, Justus Liebig.[3] Liebig had isolated a kind of fermentationlike reaction which involved crushing the hulls of almonds and adding the crushed material to almond oil. Close microscopic investigation of this reaction showed no presence of biological entities. Thus the reaction, even though it was rather like a fermentation in some respects, was even more like the typical chemical reaction in which two inert substances are merely mixed together, with a subsequent chemical interaction. Liebig concluded from this study that ferments themselves (that is, yeasts) were merely unstable

[3]Justus Liebig, *Organic Chemistry in its Application to Agriculture and Physiology* (Cambridge: J. Owen, 1841).

chemical compounds and that fermentations were self-propagating instability reactions, which occurred when the ferment was added to a suitable chemical solution.

Liebig's contemporary Berzelius had gone on to make the ultimate metaphysical claim that the fermentation reactions were produced by direct mechanical contact between the unstable ferment and the receptive structure. This model was rather like a "domino theory" in that the receptive solution was conceived to be a delicately balanced structure which was easily collapsed when it was impacted by the motion of the unstable ferment.

It is fairly plain to see what research programs were engendered by this paradigm. Chemists, whatever their specific interests, became engaged in analysis of the underlying physical structure of the compounds they were interested in. All reactions were to be ultimately analyzed in terms of the geometry and motions of the reactants. This paradigm was totally accepted and, indeed, is the explicit motivator of Pasteur's first series of investigations into chemical structure.

CRYSTALS AND LIGHT

Pasteur's first work as a professional chemist dealt with some special optical phenomena. Polarizing crystals had been discovered only recently, and much work was going on in an effort to use the new phenomena as a research tool for discovering chemical structure. Pasteur succeeded in his efforts. He found a rich vein to mine using the new polarizer tool. Perhaps I had better say a few things about how polarizers work and are used; even though we are all familiar with the effects of Polaroid lenses, their functioning processes are not so well known. First off, consider light. Light has several interpretations, several models, which attempt to describe it as a substantial object. None of these—whether we study the light "ray," the light "particle" (photon), or the light "wave"—is entirely satisfactory in all fields of optics. However, for our present purposes, let us suppose that light is an object which travels as a wave of some definite dimension. It is curvy, with a specific amount of curve per unit length, depending upon its "color" (or, technically, its frequency). Light, as it radiates out from a source, is generally in no special arrangement. The curvy waves are randomly splattered about at all angles, some vibrating from up to down, some vibrating from right to left, with others to be found at all angles in between.

The direction of vibration is called the *polarization* of the light wave. If a whole group of waves were to be radiated from a source in such a way that the curves all vibrated from the top to bottom of this page, as in Fig. 6-1, these waves would be said to be vertically polarized. If they vibrated

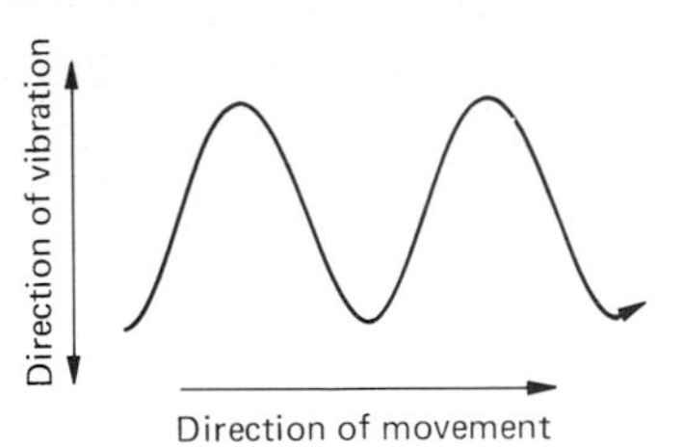

Figure 6-1. A model light wave.

in such a way that the curves stuck out from the side of the page, they would be said to be horizontally polarized. So much for light itself. A Polaroid (polarizing) crystal or lens is one which has a molecular structure arranged somewhat along the lines of a Venetian blind, or a set of shutters. Imagine a lens set up with its molecular "shutters" organized up and down, or, to be technical, with vertical polarization. If you were to hold this lens between your eye and a source of randomly polarized light, the "shutters" would allow through only that light which vibrated within the slots of the shutters. Thus, horizontally polarized light would not be able to pass through the lens simply because its curves vibrated *across* the openings of the shutters. This explanation should help you to see how polarized sunglasses work. Most of the environmental light we encounter is polarized, especially light coming directly from the sky. The Polaroid lenses screen out a certain amount of this light, simply because they function like invisible venetian blinds.

Let me now tell you how to perform a revealing experiment with Polaroid lenses. (See Fig. 6-2.) Take your sunglasses and break them into two pieces. (I realize that this is an expensive step. However, scientific truth is never got cheaply. If you prefer, you might borrow another pair of glasses and use one lens from each pair.) Now hold the two lenses between you and a light source, so that you are looking through both lenses at the light source. Rotate the lens nearest you. At some point in the rotation, the light should go dark. What has happened? It is not terribly hard to figure out: The lens nearest the light has filtered out all the randomly polarized light, and only light which has a polarization identical to the lens can pass. When you rotate the second lens, you are rotating its own plane of polarization. Sooner or later, the polarized light passing through the first lens is going to be directly across the polarization of the second lens. The polarizations are then at cross-purposes and nothing at all gets through the set of lenses. (If this does not happen when you rotate

Figure 6-2. Horizontal versus vertical polarization.

your two sunglass lenses, then the lenses are not really polarized but only some cheap imitation. As I noted above, scientific truth does not come cheaply.)

The experiments which Pasteur carried out used a set of lenses just like polarized sunglass lenses. The first lens, the one nearest the light, is called the *polarizer*, since it functions to orient all the light waves in the same polarization. The second lens is called the *analyzer*, since it analyzes what has happened to the polarization of the light. To see how this works, consider the fact that light does not pass through the set of lenses when their polarizations are at 90° from one another. But it does pass through at other angles, measured in reference to the 90° point. Suppose we start out with the lens rotated to the darkened point. If we backtrack 90°, then we are at 0° difference between polarizer and analyzer. If we stick something in between the two lenses, and that "something" itself functions as a polarizer, then the analyzer will have to be rotated a certain number of degrees off the 0° point in order to account for the second polarization which the light picked up in the thing between the two lenses. I know this all sounds rather obtuse, but it really is not obtuse. If we stick a sugar/water solution between polarizer and analyzer, the light no longer passes through the analyzer in full strength. The light does not quite fit through the polarizing screen in the analyzer. But if the analyzer is slowly rotated counterclockwise, the light will suddenly come back up to full strength. The conclusion must be that the sugar/water solution is itself acting as a second polarizer, and is rotating the plane of the vibrations of the passing light waves coming through the first polarizer. (See Fig. 6-3.) Substances which rotate the plane of polarized light are said to be *optically active.*

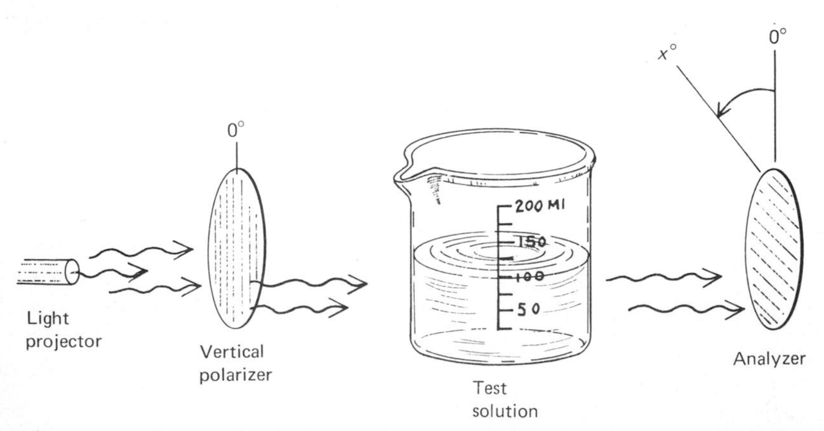

Figure 6-3. Setup of polarizer-analyzer method for investigation of solutions. This method is precisely the one followed by Pasteur. The angle x^0 is a measure of how much polarization is produced in the test solution.

Pasteur's earliest work, soon after he graduated from the Ecole, had been concerned with noting the various degrees of optical activity exhibited by all sorts of substances. In this work he had formulated a concept which he called the "law of hemihedral correlation." This "law" asserted that there was a correlation between the unsymmetrical crystalline structure of certain substances and the fact that they were optically active. This phase of his work is clearly physical, involving only the laws of optics and information about the crystalline structure of some special substances. However, during this work he had filed away a bit of data which later came to be of crucial importance, even though he neither recognized it as such at the time, nor especially concerned himself in its investigation. What he had found was that substances which were optically active usually had some connection with life; that is, they were "organic" compounds. But since this piece of data was not central to his then-current physical researches in any way, only later did it acquire its crucial meaning, and then within the new paradigm which he himself created. (An interesting sidelight is that his work on crystals itself had a conceptual guide, namely his general hypothesis that the internal, microscopic structure of substances is causally related to their optical activity. This is quite in line with his paradigm beliefs. As I said earlier, empirical research always presupposes some concept as leader and guide.)

As his work on optically active substances proceeded he came across amyl alcohol, which is a mixture of alcohols produced as by-products of the souring of milk. Amyl alcohol apparently violated the law of hemihedral correlation, and he became immensely curious about the reason for the violation. Thus, discovering all that he could about the nature and production of amyl alcohol became a primary interest for him. Since amyl alcohol was most commonly available as a by-product of the souring of milk, he decided that he also had to study the nature of *that* reaction. As we shall see in just a moment, the souring of milk was one of the keys to the whole discovery. First I must comment on what some might call the "accidental" path which led Pasteur to his study of the souring of milk.

There is no accident involved in Pasteur's ending up studying the sour-milk phenomenon. All the choices along the path from polarized light to sour milk are clearly linked together. The study of optical activity leads to investigation of many substances, and to postulation of the hypothesis about hemihedral correlation. Further research in an effort to verify this law leads to the observation that amyl alcohol apparently violates the law. Focusing upon amyl alcohol leads to research into the souring of milk. And this research will provide a key to Pasteur's discovery. There seems to be no puzzle about how Pasteur was led each step of the way. If there is any accidental feature about the whole process, if there is any surprise at where he ended up, it is simply because one would not have been able initially to predict that the physicist would end up scrutinizing a purely organic phenomenon such as milk turning sour. Pasteur himself admits, "It may seem surprising that I should take up a subject dealing with physiologic chemistry apparently quite remote from my first labors; nevertheless it is very directly related to them."[4] But as I have outlined it above, there are no surprises involved, except in the mind of someone who did not know the full story.

The feature of Pasteur's discovery which seems to best qualify as the nonparadigm element involves something he was doing at the same time as his optical work. Even while he was investigating amyl alcohol, he had been employed by the big breweries of Lille as a researcher. The brewers had been having problems with their fermentations going bad, and the beer subsequently spoiling. Pasteur was called in to see if he could not figure out some way to fix the problem. He soon found an answer, which is clearly important, but more importantly, he clarified the conceptual system involved in the brewing art. That is to say, he came up with a theory which could be linked up with the 'if, then' correlational scheme, the cookbook recipes, which had been followed by brewers since time immemorial. I will describe some aspects of these recipes and show how Pasteur related them together.

BEER AND WINE

Brewing has been a human occupation since prehistoric times. In fact, the earliest specimen we have of human writing is a beer recipe written on a clay tablet in the script of ancient Babylon. But beer making is actually a very complicated process from a technical point of view. Grain must be grown, germinated, malted, and extracted in hot water. Then yeast must be added to the extract, and the mixture must be fermented. Even when the final product is in hand, it is fairly unstable; it has a decided tendency to spoil. The crucial step in the whole process, however, and it is a point known well by even the most ancient brewers, is the step taken after the hot-water extraction: the yeast addition. Oftentimes, in an apparently spontaneous fashion, the sweet extract will take off in some sort of fermentation different from the hoped-for alcoholic fermentation. Brewers had discovered by historical times that their product benefited from the use of cool locations such as cellars for brewing and, more especially, from the addition of large masses of yeast saved up from their previous successful brews. If they followed these routines, they could be somewhat assured of not producing some foul-smelling, undrinkable mess.

[4]Conant, op. cit., pp. 22–23.

Other features had also become well known by Pasteur's time. For example, the fact that fermentations gave off carbon dioxide gas (the "fixed air" of Lavoisier's time) had been well verified. Moreover, it was known through the work of Gay-Lussac (1833) that the essential active ingredient in fermentation was the sugar. This point seemed, in one respect but not in another, to square with what was known about wine making. The similarity was that, just as the hot-water extract of malt was loaded with fermentable sugar, so also was the juice of crushed grapes. Thus the Gay-Lussac description of fermentation as the reaction

$$\text{Sugar} \overset{\text{(Ferments)}}{\Longrightarrow} \text{alcohol} + \text{carbon dioxide gas}$$

could be applied to both wine making and beer making. But in contrast to this similarity there was one very divergent point between the two processes, and as you might expect, this was the significant point for Pasteur's later results.

The processes of beer making and wine making are distinct in the fact that reliable beer can only be produced by the addition of yeast from a previous good batch. But wine can always be produced reliably with no addition of yeast. In the famous French wine regions of Bordeaux, Burgundy, the Rhone, and so on, yeast was never added to the crushed grapes. Rather they were simply set in large vats and left to themselves. Such a process would be disastrous in beer making. Because of this discrepancy between the two processes, there was a great deal of controversy about the role of the yeast. Moreover, there was some debate about what yeast itself actually was, although microscopic researches in the decade before Pasteur had provided fairly secure evidence that the yeast was some sort of a living being. Pasteur himself had decided that yeast was a species of microscopic plant, a concept which figured largely in his later discovery.

We are now in a position to see just what Pasteur learned from his research in the brewing vats of Lille. Pasteur discovered that fermentation produced a massive increase in the amount of yeast present in the extract. Indeed, sometimes there was five times as much yeast remaining after the reaction as there had been to start. He focused his microscope sharply upon the extract during the whole cycle from first yeast addition to final decanting of the beer. It was plain to see that the yeast colony went into very vigorous growth and agitated behavior in the first stages of the fermentation reaction. The absolute numbers of yeast grew in a rapidly accelerating amount. However, after this first stage of wild growth, they settled down into a roughly steady state of growth and death, with the colony's population stabilized at some final figure. So it went until all the remaining sugar was gone from the solution. At that point, the yeast activity almost ceased; the yeast apparently went dormant and fell out from suspension in the new beer, then collected in large masses on the bottom of the vat.

These observations were too much for Pasteur to handle in terms of his purely chemical paradigm. He reached the conclusion that the correlation between yeast behavior and alcohol production was too close for it not to be a necessary relation. That is, he came to view the life processes of the yeast as somehow *necessary* in the production of alcohol. This was a highly controversial point in terms of his learned paradigm, which must necessarily view the living yeast as merely an accidental component of the fermentation. In any case, Pasteur's suspicion, like all hunch concepts, had several implications which could be immediately seen. In the first place, if the yeast's life processes were a necessary element in the fermentation, then why was yeast not required as an additive in wine making? As a second point, if fermentation were related specifically to a yeast, what went wrong when fermentations went bad? Pasteur several years later answered the first of these queries when he showed that brewer's yeast cells naturally adhered to the skins of grapes even as they were ripening in the vineyard. Thus, wine makers did not have to add yeast simply because the yeasts were already present on the grapes themselves.

Pasteur provided the answer to the second question even more immediately when he pointed out that souring fermentations showed a correlated increase in the presence of other, non-brewer's yeast, microorganisms. The spoilage could be viewed as a "fermentation" but one which was linked necessarily to a microorganism *other* than the desired yeast. To put it in blunt but prophetic words, the extract had been infected by some microorganism, that is, it had caught a disease. As Pasteur was himself to say later, "Beer catches diseases, and these diseases are specifically related to the type of microorganism which gets into the beer." This answers the second question.

Pasteur's reasoning and conclusion in the case can now be stated fairly succinctly: Alcoholic fermentation in beer is the necessary consequence of the life activities of microorganisms. If there were no microorganisms, then there would be no fermentation. This conclusion is based upon observations of the life history of the yeast colony in any brewer's vat: Whenever the fermentation speeds up, so does the yeast population; when it slows down, so also does the colony; and so on. So much for what Pasteur learned from his researchers on brewing. Now let us return to his parallel work on the souring of milk.

THE LACTIC ACID REACTION

A main feature of milk souring is the production of lactic acid. This soluble crystalline acid had been isolated from sour milk eighty years before Pasteur by the Swedish chemist Schele (who is famous as one of the codiscoverers of oxygen, along with Lavoisier). Its production had been studied, and had been made much more efficient. Indeed, production of lactic acid had been shown to be most efficient when it did not involve milk at all, but rather started out from a sugar-water solution which contained some protein material and ground-up chalk. That the production of lactic acid from sugar water bore certain similarities to the production of alcohol from sugar water had not gone unnoticed. Moreover, Pasteur himself specifically and pointedly observed that there were other general similarities: both reactions foamed, both reactions gave off considerable volumes of CO_2, and so on. Pasteur's general feeling for the analogy between the two reactions will later be seen to be important in his discovery. As a final point about the lactic acid reaction, it had been noted that the naturally occurring production of lactic acid in souring milk, and its laboratory production from sugar water, were essentially similar in all regards except reliability and efficiency, both of which were increased in the laboratory method. Thus the stage is set for Pasteur.

Here is how the story goes. Pasteur begins to investigate amyl alcohol as produced when milk sours. In the course of checking out the nature of the souring reaction, he must investigate lactic acid. Since it is a crystalline substance, he routinely submits it to the optical activity tests and discovers that it, like amyl alcohol, is optically active. But the optical activity of both these substances is very different from that of the original sugar which went into their productions, and moreover, each of their optical activities seems unrelated to any of the other substances which went into their original production reaction. Whence comes the optical activity? At this point Pasteur brings to bear the observation which I mentioned earlier: He had long noted that there seemed to be a high degree of correlation between optical activity and organic processes. Indeed, he had come to realize that there was a conditional of the form "If a substance exhibits optical activity, then it most likely has an animal or plant origin."[5] However, this correlation is not particularly of interest or value to Pasteur in terms of his learned paradigm. It is just a straightforward "fact" to collect as he proceeds along his way investigating the "law of hemihedral correlation." But given the results of his work with brewing and alcoholic fermentation, the collision between this conditional and that

[5]Ibid., p. 24.

fermentation work is unavoidable: If optical activity implies animal or plant origin, then lactic acid's optical activity involves animal or plant origin. Further, if alcoholic fermentation is a reaction necessarily involving a living microorganism, then perhaps the lactic acid reaction is also a type of fermentation, which would imply that it necessarily involves a living organism. The optical activity of lactic acid would thus be explained.

This sounds like a terribly complicated piece of reasoning, and it is indeed complicated. What Pasteur has done is to take two apparently completely unrelated conceptual structures and put them together. I think that it is safe to say that this welding of the two conceptual systems, the physics of optical activity and the biologics of brewing, is the starting point of modern biology. It is clear that Pasteur's belief in the necessary connection among optical activity, life, and fermentations is just the connection needed to show that, although life involves physical phenomena, it has phenomena of its own which can provide the basis for an independent system of explanatory concepts. Thus, on Pasteur's account, fermentations can be explained in terms of living systems; there is no need to push into ever-deeper levels of explanatory concepts, concepts which ultimately will involve the underlying geometry and molecular motion of the "chemistry" of fermentations. This notion of an "independent" biology, that is, a system of explanatory concepts which is in some sense internally complete, is a controversial one, which I must say a few things about, after wrapping up the discovery phase of the Pasteur case.

THE DISCOVERY IS ANNOUNCED: FERMENTATION REQUIRES LIFE

Pasteur did not wait long to publicly declare his hypothesis. In 1857 he read a paper on the lactic acid reaction before the Scientific Society of Lille. In this paper he noted publicly for the very first time his belief in the connection between life and optical activity. Then, following this statement, he went on to make two very distinct claims (although he did not state them quite this explicitly): (1) Alcoholic fermentations are necessary consequences of the activities of living systems, namely, yeasts; (2) lactic acid reactions are in fact highly similar to alcoholic fermentations.[6] The deductive consequences of the conjunction of these two claims include one logically solid prediction, and Pasteur spent most of his effort in the paper attempting to verify this prediction. The prediction is: If the lactic acid reaction is highly similar to alcoholic fermentation, then it is the necessary consequence of the activity of a living system. From this it follows that, if a living system, a microorganism, could be found to be

[6]Ibid., p. 32.

associated with the lactic acid reaction, this would offer sound evidence that the alcoholic and lactic acid reactions are of identical types. Moreover, this finding itself will offer sound evidence for a generalization which also is a consequence, although an *inductive* one this time, of the conjunction of the two claims above: All fermentations necessarily involve the activities of living systems. Pasteur has this generalization well in mind.[7]

It is easy to see the complexity involved in Pasteur's discovery. Yet, even given the complexity, the logic can be sorted out quite clearly, and placed in a reasonable order. But first, I must give you some brief warnings.

Although it is a relatively straightforward job to re-create and reconstruct the reasoning of Pasteur in his 1857 paper, several warnings must be issued about such a reconstruction. As I noted in the Lavoisier case, it is difficult at best to discover what *really* went on in a scientist's mind during the creative act. That, however, may be a job for the psychologist, if it is a job for anyone at all. Moreover, the evidence provided by a public research paper is most likely not entirely germane to the sudden leap of insight. However, this evidence is of utmost interest for the theorist of science, who is attempting to point to the *rational* elements involved in discovery. The public research paper is the preliminary product of the rational aspect of discovery. Such papers are usually given early in the career of a theory and ultimately must represent scientists' first efforts to make their own course of thought reasonable both to themselves and to their colleagues. In this regard, Pasteur's research papers are exemplary. His logic is clean and quite clear. He is not at all hesitant to point out where he has patched up a gap in evidence with a hunch, or with a generalization which has very little evidence. Pasteur's papers are models of the processes of rational thought in discovery; in a way, Pasteur seems often to be more swayed by the deductive implications of his theorizing than he is by the actual evidence on hand. But that seems to me to be the way of most good scientific discoveries: The theory leads inexorably on, leaving the accumulation of evidence far behind. On these purely historical grounds alone, the distinction between the discovery and verification stages of scientific theorizing appears well justified.

[7]It is with certain delight that I recall to you the logically similar move made by Lavoisier: From only two cases, those of phosphorus and sulfur, Lavoisier lept to the universal generalization that "*all* weight-gaining combustions involve combinations." Pasteur's leap to "*all* fermentations involve microorganisms" was even more shaky, since it was still doubtful whether the lactic acid reaction deserved to be called a "fermentation"—although Pasteur had no doubts about the issue. I will discuss this issue in greater detail in just a moment.

A final warning must be issued. Even though clean and clear, Pasteur's reasoning is not always explicit in its entirety. We should expect this, since he is producing a scientific paper, and not an exercise in formal logic. Because of this, however, there are certain elements in his case, indeed in *any* reconstruction, which are hypothetical, and may have more of the creativity of the logician/historian about them than they do of the creativity of the scientist under consideration. The following reconstruction is guilty of this sin. But the evil is not heinous: Every single proposition is noted in some way or another by Pasteur himself in his report. Thus, if there is any change, it is not in the facts, but rather in the explicit function of these facts in the reconstruction as a whole.

Pasteur's Reasoning

Pasteur leads off with the sweeping generalization that has been with him so long that he calls it a "preconception." The generalization is the "law" of optical activity, namely, "All optically active substances involve the activities of a living system in their production."[8] This law may be understood to be the formulation of a hunch that Pasteur has because of his long series of observational correlations between organic origins and optical activity. Although we might criticize this law as being vague, unspecific, and ultimately founded on what is at best a shaky evidential basis, there is no getting around the fact that this proposition is indeed Pasteur's starting point. He himself proclaims this to be the case. But having once stated the law, he links to it the observation that lactic acid exhibits optical activity. The deductive consequences of this linkup are inescapable: If all optically active substances involve the activities of living systems in their production, and if lactic acid is optically active, then lactic acid involves the activities of living systems in its production. This inference may be expressed rather precisely in our symbolization system. We can use these abbreviations:

O__: The substance _______ is optically active.
L__: The substance _______ involves the activity of a living system in its production.
1: Lactic acid.

The inference can be laid out in a simple example of *modus ponens* (MP):

P.1	$(x)\,(\mathrm{O}x \rightarrow \mathrm{L}x)$	(1)
P.2	$\mathrm{O}1$	
C.	$/\therefore \mathrm{L}1$	(2)

[8]Ibid., p. 24.

The conclusion (2) that lactic acid production involves a living system now functions as a leading idea when it is conjoined to Pasteur's research on fermentation.

To move on to the second element of the discovery, Pasteur's close observation of the yeast colonies in the beer and beet vats of Lille had led him to make an extremely controversial generalization. On the basis of his observations he had concluded that alcoholic fermentation involved the activities of living systems, namely, the yeasts. But he did not stop there. He then generalized his results into what we might call his "law" of fermentations: "All fermentations involve the activities of living systems." The logical status of this "law," and the exact timing and ingredients of its formulation, are extremely cloudy. It simply is not known whether Pasteur formulated it *before* he came to view lactic acid production as a fermentation, or only after. In any case, the logical status of the move is very shaky. If Pasteur generalized on the basis of the beer and wine reactions alone, then he generalized inductively on an extremely shaky data base. On the other hand, if he reached his generalization on the basis of the beer and wine *plus* the lactic acid evidence, then he is in grave danger of fallacious reasoning. One of the fallacies possibly involved would be that of "begging the question," in that there are two questions involved: (1) Do fermentations necessarily involve living systems? and (2) Is the lactic acid reaction a fermentation? It is possible that many scientists would grudgingly admit that "yes" is the answer to the second question, but that they would never at the same time answer "yes" to the first question. Someone in the Liebig camp, for example, could conclude that (1) All fermentations are instability reactions, and (2) the lactic acid reaction is a fermentation. Thus, it could be claimed, Pasteur really cannot use the lactic acid evidence to shore up the conclusion "All fermentations involve the activities of living systems" until he has shown conclusively that "The lactic acid reaction *is* a fermentation." Another author has argued that the reasoning here is fallacious not because it begs the question, but because it is "circular."[9] That is, when Pasteur concludes ultimately that "The lactic acid reaction is a fermentation," he uses as part of the evidence for this conclusion the assertion that "All fermentations involve living systems"; and on the other hand, when he reaches the generalization "All fermentations involve living systems," he bases it in part on the claim that "The lactic acid reaction is a fermentation." On either of the views presented here, Pasteur's logic seems to be weak on at least one of two possible counts: On the one hand, if he *did not* use the lactic acid data, then his data base for the fermentation generalization is extremely weak; but on the other hand, if he *did* use the lactic acid

[9]Ibid., p. 35.

data, then his reasoning is fallacious by reason of either begging the question or circularity.

It is inescapable that Pasteur's logic is flawed, at least in the sense of the typical rules of logic of science. However, Pasteur seems to be aware of at least one of the problems, and thus he makes a move to at least partially deflect the logician's critical lance. As I noted earlied, Pasteur explicitly points to certain observable analogies between the alcoholic fermentation and lactic acid reactions.[10] It seems to me quite plausible that Pasteur is here attempting to weaken the thrust of the logician, at least insofar as the logician can claim that Pasteur's thought is fallacious. For this reason, that is, simply because his move makes sense when interpreted as a rebuttal to the fallacy attack, I choose to conclude that Pasteur may be guilty of the sin "generalizing from a weak data base," but that he is at least potentially justified in responding that he is *not* otherwise guilty of fallacious reasoning. The practical import of my interpretation is that we can view Pasteur's move at this stage of the discovery as proceeding independently of the data involving the claim "The lactic acid reaction is a fermentation." Thus, he will reach this assertion in an independent fashion. Let me now put all of this in a symbolic fashion.

F__: ______ production involves a fermentation reaction.
L__: ______ involves the activities of a living system in its production.
b: Beer.
w: Wine.

Pasteur took his observations of the Lille breweries, and generalized them in the following inductive fashion:

P.1 (Fb.Lb)
P.2 (Fw.Lw)
C. /∴ $(x)\,(Fx \rightarrow Lx)$ (3)

The conclusion (3) is what I earlier called the "law" of fermentations. Note that my symbolic treatment uses only the beer and wine data. Thus, it is independent of the lactic acid data. But the question arises, what bit of reasoning did Pasteur use to independently reach his conclusion about lactic acid?

Although my reconstruction is certainly only hypothetical, there is some evidence for it, and moreover, it is a most plausible version of what

[10]Ibid., p. 29.

might have gone on. Remember that Pasteur explicitly remarks that the lactic acid reaction starts from a sugar-water solution, and what is more, that it involves foaming and the liberation of CO_2 gas. These observations may be generalized in the case of alcoholic fermentations, and then, by analogy, applied to the lactic acid reaction. Here is how this two-step move probably went, as symbolized.

B__: The ____ reaction bubbles and foams.
F__: ____ production involves a fermentation reaction.
G__: The ____ reaction gives off CO_2 gas.
S__: The ____ reaction starts from a sugar-water solution.
b: Beer.
l: Lactic acid.
w: Wine.

Pasteur first generalized his observations. First move:

$$\begin{array}{ll} \text{P.1} & (((Bb \,.\, Gb) \,.\, Sb) \,.\, Fb) \\ \text{P.2} & \underline{(((Bw \,.\, Gw) \,.\, Sw) \,.\, Fw)} \\ \text{C. } /\therefore & (x)\,(((Bx \,.\, Gx) \,.\, Sx) \rightarrow Fx) \end{array} \qquad (4)$$

This conclusion (4) may be looked upon as a simple "if, then" generalization of easily observable correlations, plus their theoretical identification as "fermentations." Pasteur next uses (4) as his main premise to reach "The lactic acid reaction is a fermentation," independently of conclusion (3) that $(x)\,(Fx \rightarrow Lx)$. Second move:

$$\begin{array}{lll} \text{P.1} & (x)\,((Bx \,.\, Gx) \,.\, Sx) \rightarrow Fx & (4) \\ \text{P.2} & \underline{((Bl \,.\, Gl) \,.\, Sl)} & \\ \text{C. } /\therefore & (Fl) & (5) \end{array}$$

His conclusion (5) is one which he was ultimately interested in reaching, since now that he has reached it independently of the "law" of fermentations (3), he can use (5) and (3) together in a deduction. It goes as follows:

$$\begin{array}{lll} \text{P.1} & (x)\,(Fx \rightarrow Lx) & (3) \\ \text{P.2} & \underline{(Fl)} & (5) \\ \text{C. } /\therefore & (Ll) & (6) \end{array}$$

His conclusion here is that "Lactic acid involves the activities of a living system in its production" (6). This conclusion may now be conjoined with the very first inference he made, namely the deduction from the law of optical activity that lactic acid involves the activities of a living system in its production, which is the conclusion given as (2) above.

What Pasteur has achieved here is a marvel of scientific reasoning. He has converged on the conclusion L1 from two directions: First, the law of optical activity suggests L1 (2). Second, his research on fermentations suggests that fermented substances *also* involve living systems in their production. Moreover, his lactic acid work suggests that the lactic acid reaction is a fermentation; consequently, it involves the activities of a living system in its production [L1 (6)]. Which returns him to his initial thought that L1. You cannot miss the fact that Pasteur's reasonings have produced a tightly woven network of concepts, observations, and logical relations. Thus, what we have here is an exemplar of a theory, a conceptual system of the first order. Moreover, as I hope that I have at least somewhat illuminated for you, the reasoning which went into the invention of the conceptual system possesses a reasonable facsimile of logical cogency and coherence. And at least initially, it is this cogency and coherence which Pasteur must offer to his listeners—and to himself. As I have mentioned, he *has* to rely on the coherence and cogency of the theory to gain the attention of his colleagues, since he certainly cannot rely on the *evidence.* Once again I must point to the conceptual aspect of science as the leading edge of discovery. Pasteur's thought leapt far ahead of his practice and evidence. But Pasteur's busy times are just beginning as we must leave him. I will soon chronicle his struggle against the adversities thrown his way by the entrenched establishment, those practitioners from the tradition of the chemical paradigm.

THEORY INDEPENDENCE VERSUS THEORY REDUCIBILITY

Let me now take up the issue of the "independence" of scientific theories. I have claimed that Pasteur's work constitutes a kind of "declaration of independence" of biology from chemistry. What are the implications of this metaphor? In the first place, it has one very practical consequence, namely, that universities must now set up departments of biology. But this practical consequence, although it is extremely important, is simply a ramification of the more logical aspects of scientific theory independence. To put this in its simplest terms, biology becomes logically independent of chemistry when the 'if, then' conditionals, the empirical recipes of some technique such as brewing, come to be explained causally by concepts which do not require the use of chemical concepts. This is a somewhat simple-minded statement of what I mean, and I am sure that many of my fellow students of science will object strenuously to my saying it this way. My statement at least has the saving grace of being plain and clear as well as being possibly accurate, but I realize it may need further explanation and so will amplify upon it below.

Explanatory Closure

Pasteur's chemical paradigm required him to explain phenomena ultimately in terms of the geometrical/physical structure of the fundamental atoms, in conjunction with their motions such as dissociation and combination. Thus, Liebig's talk of "unstable structures" and Berzelius's notion of "contact action" are prime examples of chemical explanations of fermentation reactions. Pasteur, however, does not rely upon these concepts, even though they are logically necessitated by the chemical paradigm. Rather, Pasteur relies upon concepts from an entirely different system of thought, that involving life and its processes. Thus, we see him causally depending upon ideas like "living," "breathing," "eating," "excreting," and "dying" (although, of course, he uses more technical terminology than this). These concepts are connected together, and then the network is itself linked up to the "if, then" conditionals of the empirical recipes. For example, we might find the "if, then" relation "If you want alcohol, then combine yeast and sugar water" causally connected to the proposition "yeasts eat sugar and excrete alcohol." In some strong sense of the word, this explanation is "closed." There is no need to go further into the subject in chemical terms, that is, there is no necessity to begin investigating the molecular structure and atomic movement of the yeast's underlying fundamental constitution in order to account for the "if, then" statement. A main feature of this explanatory "closure" comes from two related points. In the first place, the explanation of fermentation via living yeasts is a sufficient one. That is to say, if it is true that "Yeasts eat sugar and excrete alcohol," then we know full well why "If you want alcohol, then combine sugar water and yeasts" is true. Now it is absolutely clear that we can ask a *further* question, namely, "How/why do yeasts eat sugar and excrete alcohol?" and this question might very well get us into chemical explanations. But there is a vital flaw in viewing this further situation as though it indicated that biology were not independent of chemistry. The flaw is this: Our movement beyond the yeast is an attempt to explain the *yeast*; it is *not* an attempt, in any straightforward sense at all, to explain the original "if, then" conditionals. To ask "Why/how do yeasts do this?" is not to ask the question "Why/how does combining yeast and sugar-water correlate with alcohol production?" Indeed, the "why/how" yeast question is logically dependent upon the "yeasts eat . . ." answer. These logical points about explanatory closure are linked to the other aspect of the independence of theories. Science has two main goals, namely, practical control and intellectual satisfaction. Explanatory closure seems to me to relate to the second of these.

Practical Closure

But explanatory concepts also have their practical implications. Thus, a closed explanatory system must also be "closed" (independent) in a practical sense. Pasteur's work illustrates this feature well. Once Pasteur had postulated the connection between fermentation and living systems, an immense range of practical questions was immediately opened up, and these practical questions involved practical chemistry not in the least. Thus, for example, there is the practical question "If fermentations proceed because yeasts eat sugar water and excrete alcohol, are there any practical ways in which this activity might be enhanced?" Pasteur immediately sees one way, and hypothesizes that protein material might be a nutrient for the yeast. In effect, as he realizes, the protein functions as does a "fertilizer" (obviously, he calls it "manure") in growing a crop.[11] In our modern language, which is not all that far from Pasteur's ideas, the protein material functions like a dose of "vitamins and minerals" for the bugs. Another practical question is immediately raised as well: "If yeasts are living, then they must be 'breathing'; how does this occur?" Pasteur rapidly discovers that yeasts do indeed breathe, and in fact, alcohol production is enhanced when the yeast's oxygen supply is cut off. Since they can no longer get it from the environment as a gas, the yeasts produce oxygen from the sugar—and this process in turn produces the alcohol. As I shall claim later on in my discussion of the verification phase of Pasteur's work, it is the immense range of practical possibilities which to a great extent assures his new conceptual system of a wide and intense initial investigation.

Thus, there seem to be two senses of "explanatory independence," one related to the logical closure notion that a set of explanatory concepts is closed when it is sufficient to causally explain a set of 'if, then' conditionals, and the second related to the more practical aspect that a set of explanatory concepts is independent when it has immediate practical ramifications entirely and intrinsically its own, ramifications which may be investigated independently of any other competing, apparently logically connected, system of concepts. Pasteur's theory about the connection between life and fermentation, the main element of the first stages of the paradigm we now call biology, is just such a set of independent concepts with regard to the then-contemporary system called chemistry.

The Response to All This

Even given this clarification of what I mean by "independence," there is still the probability that some of my colleagues will take serious exception to my claim for the independence of biology. Their objection runs like

[11]Ibid., pp. 28–29.

this: "Any idiot knows that living systems are *really* composed of cells, and cells are *really* composed of chemicals, and chemicals are *really* composed of physical particles. Thus, biology is not independent of physics; indeed, there *really* are no independent objects or processes except the fundamental ones we talk about in our physics theories."

My first response to this would be to shrug my shoulders and ask "So what?" Then I would just repeat what I said above about independence, and ask my opponents if my claims are not true, i.e., if it is not true that biology is independent in just the logical and practical senses I outline. It seems to me that my claim is undeniable. However, they would then come back at me with *their* claim about what is *really* the case, namely, that physics is the most fundamental theory and thus it talks about the most fundamental (the most real?) sorts of things. And what would, what *could* I say to that?

My first response would be to note that what we are faced with here is a metaphysical scheme, a proposition which is about what kinds of things are really real. In the first chapter, I called this particular proposition "reductionism" because what it claims is that the reality of biological objects is reducible to the reality of physical objects. But this seems to me to be just plain false, at least in one stubbornly literal sense. It is just plain false that biological objects (and let me here take some rather special biological objects, namely pets and other humans, as my example) are not *really* real. That is, they really *are* really real. In fact, in terms of our practical, emotional, and most likely, our intellectual life, these biological objects are infinitely more real than the atoms and the void of physics. Although this might seem a silly response to the metaphysical claim being advanced against my view, I just want to make extremely clear what it is that we are really talking about. The "really real" of a metaphysical scheme is obviously quite different from the "really real" of our everyday lives. Insofar as the reductionistic scheme's "really real" tends to get mixed up with the "really real" of everyday life, it is pernicious. As some of my moral-philosopher friends have claimed, "Even though people might be really (metaphysically "really") just atoms and void space, this does not mean that we really (the real "really") ought to treat them that way. It is simply false that people are really just atoms and the void." So let us be clear that the sense of "really" which is metaphysical is really different from the sense of "really" which obtains in our everyday encounters with people and other animals.

All this having been said, I am still confronted with the now purified sense in which "Biological objects are really (metaphysically) just physical objects." By this point in the discussion, however, I am quite willing to accept this proposition as being metaphysically true. That is, it seems to me that, in principle, there is some system of concepts which can provide a detailed explanation for everything in the universe, and this set of concepts probably refers to fundamentally physical objects (rather than fundamentally psychological, or economic, or biological, or other, objects). In a way, this belief (and others of a similar *logical* structure but not necessarily of similar naturalistic content) does just what it is supposed to do: It provides a certain amount of intellectual comfort and ease. That *is* after all what metaphysical schemes are supposed to do. But there is a host of things which the truth of this metaphysical proposition does *not* do. For example, it does not in any way imply that biologists are going to be put out of work, or else they must be retrained as physicists. Similarly, biology texts are not to be recalled in exchange for physics texts. Probably this will never happen, even if the reductionist metaphysics *is* true. There are several reasons for this. The first relates back to my initial discussion about independence, where I claimed that logical sufficiency of explanatory concepts was associated with the raising of practical possibilities. Thus, independent and closed theories have practical techniques inherently and essentially connected with them. Biology clearly exhibits this feature. This is a positive reason that biologists will continue to have employment. Another, but negative, reason also exists. It involves explicit practical difficulties.

Physicists Are Not Biologists

The reductionist thesis implies that chemists will lose their jobs to physicists, even before biologists will lose theirs. I had a friend in graduate school who actually believed a weak version of this, and attempted to carry it out in an experiment which eventually grew into his doctoral thesis. This fellow was a chemist who believed that reductionism was the true metaphysics. Thus, he reasoned, chemistry should be do-able using physical concepts, physical theories. After all, chemical compounds are *really* (there is that metaphysical "really" again!) only physical particles in certain arrangements. Consequently, he reasoned, it should be feasible to describe a chemical compound in strictly physics terms. So what he did was to take the methane molecule, a very simple chemical substance which involves only five atoms, and attempt to describe it in today's most modern physical theory, the quantum theory. Now quantum theory is tough, really tough, and it involves horrendous statistical formulae which march across computer program pages, the way huge, never-ending German nouns march across the pages of existentialist philosophy books. After three years of work, cutting, pasting, coming up with incredible statistical approximations (one was so long that it managed to overflow the memory of the campus computer, which then promptly shut down), sleepless nights and so on, he produced a quantum theoretical model of the methane molecule which was sufficiently close to be acceptable. (However, he readily admitted that it was not *really* all that close; his advisors had felt sorry for him and let him go.) By this time he had come to the conclusion that, even though the reductionist thesis might very well be true, it was not really a very *interesting* thesis. To end the story, I understand that he is even today a practicing chemist, but he does not talk much about physics any more.

The moral of the story is that this particular "in principle" truth just is not very significant. Indeed, except in the pernicious sense I decried above, the reductionistic thesis seems relatively innocuous, indeed, harmless and comfort-inducing.

I wish to end this discussion of theory independence with a point once made by my good friend, psychologist Bob Sanders. Sanders has pointed out that some theories remain independent—and probably will remain eternally so—because, to put it bluntly, they have become "finished" or "perfected." That is, there is no more work left to be done on them. Let me give an example. Galilean/Newtonian mechanics is a perfected theory. There just is not anything further, there are no remaining "secrets," left to discover about the motions and forces of moving objects. One solid indication of this is the fact that no one is now being graduated from university with a research specialty in the Galilean/Newtonian laws of motion. There simply *is* no further research needing to be done. Moreover, and this is an important point, no one even bothers to attempt to reduce the Newtonian explanations to ones given in the more modern terms of Einstein's special theory of relativity, which is the theory more logically fundamental than classical Newtonian mechanics. Now this is not to say that it *could not* be done. Everyone agrees that it is in principle possible to reduce Newtonian mechanical explanations in all cases to Einsteinian relativistic ones. But who in the world would want to do it? Indeed, *why* in the world would anyone want to do it? The Newtonian explanation is elegant, simple, true enough, adequate enough for any conceivable purpose in the cases it applies to, and so on. Thus, as in the methane case above, to say that the one explanation is *reducible in principle* to the other is perhaps true, but it certainly is of no apparent interest to anyone, at least in any significant sense. It seems to me that "in principle" arguments, aside from their possible pernicious effects (which, I must emphasize, are real and dangerous, even though they almost always arise from some sort of confusion), ultimately collapse under the force of practical problems. And in their collapse, they take with them any real ramifications and interest they might have.

I seem to have strayed quite some way from my initial purpose, which was simply to explain how it was that Pasteur's work constituted biology's declaration of independence from chemistry. However, even given my rambling path, I should think that what I intend by "independence" is now a little clearer. But whether it is or not, let us now turn our attention to another (our final) case history of scientific discovery, Pauli's hypothesis that there exists a subatomic particle called the "neutrino."

SUGGESTIONS FOR FURTHER READING

If you are interested in reading some more about Pasteur, his biography and some of his essays are to be found in Hilaire Cuny, *Louis Pasteur* (New York: Paul S. Erickson, Inc., 1966). The essentials of wine making are found in two essays, F. Drawert, "Winemaking as a Biotechnological Sequence," and A. Dinsmoor Webb, "The Chemistry of Home Winemaking," A. Dinsmoor Webb (ed.), *Chemistry of Winemaking* (Washington: American Chemical Society, 1974). The best source on brewing—although it is quite technical—is J. S. Hough et al., *Malting and Brewing Science* (London: Chapman & Hall, 1971). Reductionism, as a philosophical issue, has occupied much time and space for many philosophers. The standard account is probably Ernest Nagel, *The Structure of Science*, (New York: Harcourt, Brace, and World, Inc., 1961); see especially pp. 429–446. For a differing account, see Marjorie Grene, "Hobbes and the Modern Mind," in Marjorie Grene (ed.), *The Anatomy of Knowledge* (Amherst: The University of Massachusetts Press, 1969). This essay presents cogently some dangers of reductionistic thinking. Grene's essay "Reducibility: Another Side Issue" gives some nice counterarguments to the reductionist view. It appears in Marjorie Grene (ed.), *Interpretations of Life and Mind* (New York: Humanities Press, 1971).

Chapter 7

Pauli and the Neutrino

INTRODUCTION

We now confront one of the most fascinating cases of discovery in modern science, a discovery which might not actually count as a discovery at all: the "discovery" of the subatomic particle called the *neutrino*. My telling of the tale of the actual events of the discovery will not be long or involved; but my attempt to analyze out the *logic* of the discovery from the tangled skein of events will be an arduous task. There are several reasons for this. In the first place, as you well know, my thesis is that discovery proceeds with the mind leading the hand (or more precisely, theory leading practice) to the ultimate recognition and identification of the discovered entity. This process quite clearly occurred in the neutrino case, as I shall soon show. But there are some curious immediate effects of this particular case of conceptual guidance, and these are what bring in the troublesome aspects of the situation. In the two earlier cases, it was fairly simple to identify just what were the conceptual elements leading and guiding the scientists. Moreover, as we saw, in both cases concepts *not* based on the prevailing paradigm became intimately involved in the creative act of the discovery especially insofar as they caused friction with paradigm ideas. But our present case does not go quite the same way. In the present case we shall see that friction arises, not between paradigm concepts and nonparadigm concepts, but rather between two paradigm concepts, one on a "deeper"—more fundamental—level than the other. To put it in another way, the investigation itself produces results which conflict with principles that are more fundamental in the paradigm. In this regard, since it does not involve collision between two distinct conceptual systems, we can plausibly call the neutrino discovery a *conservative* discovery, in order to distinguish it from the more radical kinds of discoveries made by Lavoisier and Pasteur. In a sense Lavoisier and Pasteur made discoveries which were radical because they created conceptual revolutions *against* prevailing paradigms. Lavoisier's chemical revolution, for example, provided the fundamental concepts for the beginnings of an entirely new chemical tradition. Thus, there was still a "chemistry" after Lavoisier, but it was a revolutionarily new chemistry as compared with what had gone on before. Pasteur's revolution was even more radical; a science called "biology" existed after Pasteur, whereas there had been no such thing prior to him.

The neutrino discovery produces no such radical effects; there is no revolution which does away with a prevailing paradigm. Rather, what happens is that there is for a time extreme logical and personal tension and conflict between both the concepts and the scientists involved in the prevailing paradigm. But after a period of this friction, things apparently settle down, and what finally turns out is that the old paradigm accepts the new theory—the new conceptual structure—to shelter within its conceptual arms. The new theory, however, even though it eventually *does* become logically meshed into the paradigm, requires modifications to be made to the original overarching conceptual system. These changes are serious, but not ultimately revolutionary. Because of this, the discovery and its changes, no matter how novel, do not "overthrow" the earlier paradigm; thus this case does not seem to fall under the notion of a "revolution" according to the view of Kuhn. However, even though the neutrino case is accordingly not a revolution, it still provides some extremely useful information about the structure and dynamics of paradigms. The reactions, moves, and countermoves which the physicists made reveal much about how the conceptual system of physics is actually constituted, and how it actually functions in real-life events. My analysis will bring out many features of these situations. In particular we will look at the idea of the "level" of concepts in a paradigm, eventually to focus upon the question of what "fundamental" means. Along the way we will consider the meanings of the concepts "hypothesis," "law," and "principle" as well. Another issue which will be threaded into almost every nook and cranny of my discussion is that of the conservation laws, those principles which apparently are the ultimate foundational cornerstone of physical science. These principles, I will claim, reflect a deep metaphysical commitment, a commitment without which the scientist might not be possible. My conclusions on this point will be fundamentally speculative—they could not be otherwise—but they are nonetheless necessary, since unless they, or some similar ones, are correct, science and the human drive to scientific knowledge are fundamentally inexplicable.

As you can tell by now, this present discussion will be a long and somewhat complicated one. But let me begin with something simple—the tale of the neutrino. By the late 1920s physicists had begun serious investigations into the construction of the atomic nucleus. Since Bohr's development in 1912 of the theory that atoms were composed (at least) of positively charged chunks of matter called "protons" and negatively charged chunks called "electrons," physicists had been attempting to carry out analyses of various elements in an effort to see whether these elements could be conceived to be constructed by carpentry work using the electron and proton as lumber. The hydrogen atom had been successfully described in these terms; it was believed to be a very simple structure composed of just one proton and just one electron. The proton was the "nucleus." It played a role in the hydrogen atom analogous to the role played by the sun in our solar system. The electron played the role of the earth. This simple theory worked well, and it could account for a whole series of experimental results. However, as the investigated elements got more complex, that is, as the studies moved toward heavier elements such as carbon, nitrogen, cobalt, and so on, it was discovered that the usual carpentry work became ineffective: There were holes in the structure which could not be patched over with the type of lumber provided by the electron and proton. A sense of real and growing crisis began to pervade the community of physicists. One small field in particular was in disastrous shape, and everyone knew that it was in immediate peril of conceptual collapse.

The field was called *beta decay*, and it was a small part of the general field involving radioactivity and particle physics.[1] Decay is a process in an atom which is just like it sounds: The atom comes apart at the seams, and explodes into parts. Decay processes are extremely valuable, since if the physicists can collect all the parts as the disintegration happens, they can then reason backward to figure out what the substance was like *before* the

[1]A general account of beta decay is in Harry J. Lipkin, *Beta Decay for Pedestrians* (Amsterdam: North Holland Publishing Company, 1962).

decay. In the case of atomic nuclei, decay allows a peek into internal atomic structure which is usually hidden. One physicist has compared this method of analysis to that which might go on when a watch comes apart: If we can track down all the watch's internal parts, and have a clue to their angles and directions at the instant of disintegration, then it becomes somewhat possible to reconstruct in our minds what the internal structure of the watch was, as a unified system prior to the decay.[2] This analogy is a fairly good one, except for one problem: We already know what a watch is like even before we try to retrace its exploded parts and their trajectories. But we have no clue about the nucleus, prior to trying to retrace its parts and their trajectories. This makes the job quite a bit harder.

I think that you can probably see how decay is a useful tool for the physicists. Unfortunately, one particular kind of decay, beta decay, proved to be more of a puzzle than a purely useful tool. Because of the puzzle, beta decay itself became the end, not the means, of the investigation. Beta decay is that kind of nuclear disintegration which occurs when an electron (e) is one of the parts which comes shooting out from the nucleus. (Other kinds of decays do not shoot out electrons.) Beta decay has more recently been described as just one of the many kinds of reactions which involve an entity called the *weak force*—where "weak" refers to a comparison with another force (naturally called the *strong force*) which holds together the parts of the nucleus. Although this is another story, one which we certainly do not want to go into here, physicists believe that there are only four different kinds of forces (weak, strong, gravitational, and electromagnetic) functioning in the universe. Thus, they believe that if they had a good theory for each one of these different forces, and also a theory of cosmology which could describe the interactions of stars and galaxies, then these five theories would completely explain all the physics of the universe.[3] Of course, our friend the reductionist—whom you have already seen several times before—believes that even five different theories is four too many; his ideal, and I must admit that it is an attractive ideal, is that maybe there is a *supertheory* which can encompass all the five within its logical arms. Once we had the supertheory, the unified theory, well in mind, everything knowable about physics could be known in light shed by the theory. It would just be a question of using logical deduction to come up with conclusions implied by the fundamental laws. The rationalist epistemology which you have seen from time to time is quite evident in this ideal of physical knowledge.

[2]Quoted from Richard Feynman on a "Nova" television program.

[3]Geoffrey F. Chew, Murray Gell-Mann, and Arthur H. Rosenfeld, "Strongly Interacting Particles," *Scientific American*, February 1964, pp. 74ff.

Beta decay theory by 1930 was in big trouble. Some very careful measurements had been made, and the various numbers belonging to the parts did not add up to equal the numbers belonging to the predecay wholes. Apparently the amount of energy in the reaction had not been conserved. To put this another way, the energy balances, the ledger sheets introduced into science by that long-ago master accountant Lavoisier, were violated by beta decay. The predecay object had more energy ("more" mass and momentum as well) than did the sum of its parts. Moreover, the spectrum (curve representing the wavelengths) of the beta particles had an entirely unexpected shape, one which was in drastic conflict with the going theory. What was to be done?

THE NEUTRINO HYPOTHESIS

The move to solve the problem was made by a young physicist named Wolfgang Pauli. To everyone's surprise, the solution unexpectedly came in a letter Pauli wrote to Geiger (of Geiger counter fame) for him to read at an international meeting of physicists. The meeting had been convened to discuss problems with beta decay and other nuclear processes. One delightful sidelight to this whole adventure is that the twenty-year-old Pauli wrote the letter because he did not want to come to the meeting himself; he had decided to stay in Zurich in order to go to a big dance. Pauli made a proposal that absolutely rocked physics, but he missed all the fireworks in order to keep dancing. Geiger read the letter to the assembled physicists with some anxiety. His anxiety was well founded. Pauli's proposal was met with immediate hostility and skepticism. As C. S. Wu has noted, Pauli's "outlandish" hypothesis required great courage by its inventor.[4] Pauli himself later ruefully admitted that the neutrino hypothesis had long "persecuted" him. But the hypothesis itself seemed to be both simple and elegant: The energy numbers did not balance because one of the exploded parts had been going unobserved. If this particle could be observed, and its energy numbers measured, then they could be added into the equation and balance would result.

I am sure you are wondering, "What is so outlandish about this proposal?" At first glance the proposal looks quite reasonable. But this is only at first glance. When the hypothesis is checked out further, the difficulty leaps into view. Here is the problem. First, in order to balance the electric charge equation, the new particle must have no electric charge. Second, in order to balance the mass equation, its mass must be nearly zero, or equal to zero. Third, in order to balance the energy equation, the particle must be extremely small, and travel at or near the

[4]C. S. Wu, "The Neutrino," in M. Fiery and V. F. Weiskopf (eds.), *Theoretical Physics in the Twentieth Century* (New York: Interscience Publishers, Inc., 1960), p. 250.

speed of light. Thus, this new particle must be neutral in charge, must be vanishingly light in mass and vanishingly small in size, and must travel as fast as can be. In short, the new particle must be, for all practical (indeed, for almost all theoretical) purposes, unobservable. Notice how the logic of this move works: The observable measurements indicate a violation of the conservation principles. What to do? Postulate a particle which can patch up the discrepancy. But in order for the patch to work properly, the particle must necessarily be unobservable. And maybe not just *technically* unobservable, but unobservable in principle, hidden forever no matter what new instruments are invented. The outlandishness of Pauli's proposal should now be apparent to you. Pauli has "discovered" the neutrino just in proposing it; there is nothing more to the discovery than this. But what has he discovered? Something which apparently can never be sensed by the eye or hand of man; something which can be "observed" only by the mind of man. But this, according to the tradition of science itself, simply will not do. Let me elaborate on what is going on here.

Pauli's neutrino comes drastically close to violating one of the most fundamental rules of the naturalistic epistemology of science. That epistemology requires that all knowledge be of a sort which can be verified using empirical, sensory means. The corresponding metaphysics of science requires that all real objects, all those entities which are to be accepted as existent, must be such that they can in some way interact with us and our physical extensions—our meters, dials, probes, and whatnot. Scientific epistemology and metaphysics come together at the point where many scientists reason, "If we cannot know something empirically, then it is not interacting with any of our senses or their extensions; but if it does not interact, then it is not real." If we substitute the term "neutrino" into this argument in place of "something," then the outlandishness of Pauli's hypothesis becomes even clearer. Pauli has proposed something which, by prevailing scientific philosophy, can be only imaginary. According to the philosophical rules the neutrino is an unreal object proposed to solve a very real, very observable difficulty. In the face of this, we must ask why Pauli and his proposal were not immediately laughed out of the community, why the physicists did not simply cast him out into the darkness as punishment for his indiscretion. The reason is extremely revealing of the logical structure of scientific conceptual systems.

Think back to the reason lying behind Pauli's proposal: On the one hand, the observed beta decay situation apparently violated the conservation principles. But on the other hand, if neutrinos existed then the energy balance would be safe. This, it is absolutely clear, is a sufficient reason for Pauli's hypothesis to be tentatively acceptable, however grudgingly, to his colleagues. There is, of course, an alternative: Pauli's proposal could be rejected, and the beta decay discrepancy could be accepted as a brute fact. But this move is apparently unthinkable (although some scientists, notably Enrico Fermi, apparently thought about it, however briefly) because of its logical consequence. If accepted, it implies that the conservation principles are false. The whole situation develops into an either/or dilemma: Either "If neutrino, then conservation is true, but the observability criterion is discarded" or "If no neutrino, then the observability criterion is kept but conservation is false." In either case the physicist is left with a very negative consequence. He must either give up a fundamental belief about his methodology (the observability criterion), or he must give up a fundamental belief about the ultimate construction of the world (the conservation principles). The choice seems devastating. But there was almost no hesitancy expressed by the physics community about which choice was the preferred one. Almost to a person, they voted for the conservation principle by affording a hostile, grudging, skeptical, but nonetheless real admission of Pauli's proposal to the sacred circles. They protested vigorously; indeed, a minority to this day still refuses to accord ultimate existence status—table-and-chair existence status—to the neutrino. (I will never forget my surprise when the professor in my graduate physics seminar asked for a show of hands on the question, "How many of you *really* believe that neutrinos exist?" A majority did, but that is not the important point. What was significant was the asking of the question itself. It had never occurred to me that existence questions—questions of the sort "Do you *believe* in *x*?"—could arise in science. Always before it had seemed to me that either *x* existed and we *knew* it, or it did not exist and we knew *that* too. Later I was to be even more surprised when I encountered some scientists who raised serious questions about the existence of other extremely small objects, such as electrons and quarks.)

This whole set of happenings illustrates an important point about the conceptual structure of science: Some concepts are more fundamental than other concepts, but some fundamental concepts are even more fundamental than other fundamental concepts. The conservation principle, at least according to the neutrino case, is even more fundamental than observability. But this notion of "more fundamental" is a devilishly tricky one. Logicians have wrestled with it for centuries without complete satisfaction with their results. One difficulty is that fishing for an answer to the question "What does fundamental mean?" invariably catches the unwanted concepts "law," "principle," and "hypothesis," which seem to be inextricably intertwined with the concept "fundamental."

THE CONCEPT OF FUNDAMENTALITY

The terms "law," "principle," and "hypothesis" are used in a hopelessly muddled fashion by all parties concerned with dissecting science. But the terms are also hopelessly muddled as they are used by scientists themselves. For example, some concepts are called "laws" by some people, sometimes; but these selfsame concepts are also called "principles" by some people, sometimes. And often the names are interchanged by some persons at different times. I have probably committed this sin of interchangeability within the last fifty pages. Underneath the mud and confusion, however, there are some real distinctions to be made among these three terms. That is, each term can be worthwhile in describing separate and distinct concepts. But as you will see, the entities which are separate and distinct oftentimes blur into one another over the passage of time; this tendency has been a major contributor the incredible sloppiness of the popular and technical usage of the three terms. I will try to sort them out a bit.

In the first place, whatever laws, principles, and hypotheses are, they are formulated in the logical form called *universal generalizations*. You have been repeatedly confronted with this logical form; it is none other than the universally quantified 'if, then' conditional, e.g., $(x)(Fx \rightarrow Gx)$.[5] But this logical sameness is the end of their close similarities. From this point onward the case gets messier and messier. A main reason for this is that these three concepts blend together aspects of scientific logic, scientific epistemology, and scientific psychology. The logical aspects of this blending have been most studied, and can be most easily reported. Consider the following analogy, which can provide a model of logical

[5]I must point out to you now what I have (not entirely honestly, I must admit) neglected to mention to you in the past: Very few professional theorists of science *literally* believe that the universally quantified conditional is the *real* ultimate form of scientific generalizations. This form is just too logically simple-minded to handle all the tasks which must necessarily be carried out by the scientific generalization, whatever its logical form is. For example, it does not allow for expression of cases involving *potential* features, that is, features of objects which are capacities. Thus, suppose we want to define "fragile" by saying "*x* is fragile only if it breaks when struck with a hammer." But if we used this definition, we could legitimately call *x* "fragile" only when we broke it with a hammer. This is a rather severe restriction on our ability to logically define properties such as "fragile." Given such a logical simple-mindedness of the universal conditional, you are probably right now asking—with justified anger—"Then why are you using it?" This is an excellent question. But I have an excellent (I hope!) answer: No matter what my colleagues might tell you, tell me, tell one another, in the final analysis we do not have a plain idea about what might be a comprehensible, more realistic candidate. I am not saying that we do not have some clues: We know that the subjunctive mood plays a part; we know that the universality of some generalizations is not physical, but rather is some sort of epistemological "necessity"; and so on. But, to state it in brutally frank terms, there does not yet exist a logic of the subjunctive, nor is there any professionally accredited account of the requisite "necessity." Consequently, I must fall back upon the quantified arrow as a model of generalization. But our risks in doing so are not entirely pernicious. The arrow model has the saving grace to communicate some worthwhile elements of the real case, and to do so with clarity. Indeed, it seems to me that whatever view of generalization we end up with some centuries hence, that view will contain, albeit as modified, the quantified arrow at least as a constituent element. Thus it is to this degree accurate for me to use the arrow model. And that is close enough for comfort.

structure. A physical structure can be envisaged as existing in layers of support. Suppose that we are talking about a pyramid-shaped structure. The widest part of the pyramid is the most structurally requisite: The foundational base is such that, without it, none of the rest of the structure would be possible. In logical terms, the base is a *necessary* condition for the rest of the structure; but even though necessary it is not a sufficient condition for the rest of the structure to exist, that is, the existence of a base does not absolutely mean that a tip exists too. Thus, we cannot reason from the base to the tip. However, we can go the opposite way. Thus, the tip is sufficient evidence for us to reason to the existence of the foundation base. This sort of reasoning provides a model of logical structure which we can apply to some particular paradigm in science, say, physics. For example, there are some generalizations in physics which are taken to be necessary to the existence of other generalizations. This provides a definition of *fundamental*: Those generalizations which are necessary to the existence of all others are the most fundamental. Using this definition we can say that if conservation is preferred over observability, then the physics community sees it as more fundamental, more necessary, than observability. Nicely enough, our analysis not only sorts out something about the logic of what "fundamental" means, it also gives us a handle on the relations among the other three tricky concepts; that is, we can say that principles are necessary for laws, and both principles and laws are necessary for hypotheses. Although this statement does not answer *all* the questions raised about the four concepts, it does help us pick our way through the maze of conflicting interpretations.

I must note something further about the logic of this model. Let us suppose that a hypothesis (the least necessary of the three types of generalizations) turns out false. Does this falsity imply the falsity of any more fundamental generalizations, for example, some principle or other? No it does not. We cannot reason from the falsity of a sufficient condition to the falsity of a necessary condition. In terms of our pyramid structure, we cannot reason from the falsity of the existence of the tip, to the nonexistence of the base. This move is invalid and is called "denying the antecedent." The argument goes this way: Suppose it is true that "My car runs only if the electrical system is functioning." In this statement "My car runs" is the sufficient condition; "the electrical system is functioning" is the necessary condition.[6] I can reason conclusively from the fact that my car runs to the fact that the electrical system is functioning; the car's running is sufficient evidence to validate that mental movement. But on the other hand, suppose that my car does not run; is that sufficient evidence to conclude that the electrical system is not functioning? Not

[6]Irving M. Copi, *Introduction to Logic* (New York: Macmillan Co., 1972), p. 265.

really; my carburetor may be out of whack. Thus, as this example shows, we can reason from the truth of the sufficient condition to the truth of the necessary condition, but we cannot make the same move with the falsity of the sufficient condition. In terms of the structure of paradigms, this implies that the falsity of some less fundamental hypothesis does not affect the truth or falsity of some more fundamental principle which is necessary for that hypothesis.

But now we must look at the other side of the coin, since this aspect of the logic of the model will indicate something worthwhile about the dynamics of scientific paradigms during revolutions. Let us see what kind of reasoning we can carry out between the necessary and the sufficient parts of the structure, except this time we will try to move in the opposite direction, from the necessary to the sufficient. Try the first move: Can we move from the truth of the necessary to the truth of the sufficient? No, we cannot. Again, this attempted move has a special name—the fallacy of "affirming the consequent." Here is how it works. Suppose that my car runs only if the electrical system is functioning. Suppose it is further true that the electrical system is functioning. Can I infer that the car runs? Of course not, as all car drivers know. Even though the electrical system functions, the carburetor may be out, or the head gasket blown, or some such. Hence the car will not run. On our model of a paradigm, this means that just because a fundamental principle is true, this does not imply that some less fundamental hypothesis is also true. Far from it. Thus, we cannot reason from the truth of a principle to the truth of a hypothesis. But there is a path left open to us: We can reason from the *falsity* of a necessary condition to the *falsity* of the sufficient condition. Thus, from the fact that it is false that "The electrical system is functioning," we can safely infer that it is also false that "My car runs." That is, the fact that the electrical system is out of whack implies that the car will not run, since my car runs only if the electrical system is functioning.

This last inference is revealing. When applied to a scientific paradigm, it means that if a principle is false, then so is its associated hypothesis. This is a very important point about the dynamics of paradigm revolutions. If the new paradigm, the new but opposed conceptual structure, attacks very fundamental concepts in the established system, then the wreckage is going to be much greater than if a less fundamental concept were put under siege. Although this is a very intuitive notion I mention here, it can be given quite a bit more precision when we model our definition of fundamental upon the logical model given in terms of the inferences involved in "necessary" and "sufficient."

Let me now sum up this phase of the discussion of "fundamental" and its associated ideas "principle," "law," "hypothesis." We use the model of a pyramid in Fig. 7-1. The most "fundamental" level is that of

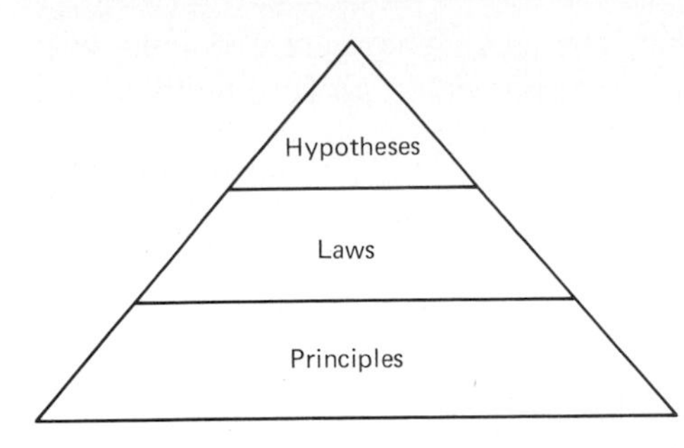

Figure 7-1. A model of logical structure.

the principle: If it fails, then all else topples with it. In logical terms, the principle is more necessary than the law, and the law is more necessary than the hypothesis. However, simply because the principle is true, this does not mean that the associated law or hypothesis is also true, since the truth of the principle is not sufficient evidence that the law is true. On the other hand, the truth of the law *is* sufficient evidence to infer the truth of the principle; but not vice versa. The upshot of the whole analysis is that the discrediting of a fundamental principle does much more damage, logically speaking, than does the discrediting of a less fundamental law or hypothesis. Given this view, we can say that Pauli and his friends felt that the conservation principle was more fundamental than the observability requirement. Unfortunately, even though this clarifies the ideas a little, problems still remain in the notions of "law," "principle," and "hypothesis" vis-à-vis fundamental.

There is another ploy logicians have tried to use to make some sense of the concepts. This strategy intends to define "fundamental" in terms of generality. Generality here has a specifically logical meaning, but it also has an intuitive connotation. One generalization is more general (more fundamental) than another if it covers more cases than the other one does. This idea is not terribly satisfactory, since it is not clear just exactly what "more" might mean. For example, a hypothesis about carbon might be more general than a hypothesis about uranium, just because there are more cases of stuff involving carbon than there are involving uranium. This seems like a trivial example, but it is the sort of objection which can be raised about the intuitive notion of "generality." But this idea can be refined a little, however, which makes it a bit more satisfactory: For example, a hypothesis about organic compounds would be less general than a hypothesis about carbon, not only because it covered fewer cases, but also because those cases covered by the organic compound hypothesis are merely a subset of the total cases covered by the carbon compound hypothesis. This idea of generality involves a sort of intuitive notion of "logically included" (rather like our earlier concept of "reductionism," where the science at the end remaining, that science to which all others are reduced, "includes" those other sciences). There are problems with *this* version as well, but I do not see any need to go further than this, which is sufficient to show that attempting to provide a completely logical analysis of the meaning of "fundamental" (and "law," "principle," and "hypothesis") certainly is fraught with peril. Perhaps we can avoid this peril if we avoid limiting our definition to purely logical features of the concept.

"More Fundamental" Means "More Cherished"

One of the other features of the concept of "fundamental" has already been hinted at. Above, in the discussion of necessary and sufficient, I noted that scientists often are very hesitant to relinquish their beliefs in the more necessary conditions of the paradigm; that is, the more fundamental the beliefs, the more dearly they are held on to. But resistance to disbelief is clearly, at least in part, a psychological feature of fundamental-concept behavior. That is, it does not logically follow from the concept of a "principle" (a proposition more necessary than some other) that it will be held on to more dearly than a "hypothesis." But it certainly follows from the conjunction of this latter concept with what we know about human nature and human behavior. It is pretty clear that if a group of scientists would have to declare false a whole set of their theories were they to accept the fact that some necessary principle is false, then they will resist having to accept the fact that the principle is false. It is a plain psychological datum that human beings tend to hang on to their belief systems. This fact provides another means of analysis of *fundamental*: If two new concepts wreak differing amounts of havoc in the mind of the scientific community, then the community will choose that belief which does the minimal amount of damage to the prevailing paradigm. Thus we have a suitable explanation of what happened in the neutrino case. If the principle of conservation were accepted as false, then an incredible amount of physics (and chemistry, etc.) would have to be redone, since the conservation principle was necessary to almost all previous results. But if the observability criterion were weakened or discarded, at least in this case, there would be less damage. Acceptance of the neutrino could be interpreted merely as an agreed-upon, moderate exception to the general requirement that "Only observable objects are acceptable to the epistemology and metaphysics of science." Thus, conservation stayed and observability went in the neutrino case. (As you will see in the section on verification, I am being overly simple here. The actual facts are a little more complicated, but this presentation is close enough to being accurate that we need not worry about it now.)

Thus we now have another criterion to use in defining *fundamental*: Concept *X* is more fundamental than concept *Y* only if the scientific community is more reluctant to disbelieve *X* than to disbelieve *Y*. On this account, principles are clung to with greater tenacity than are laws, but laws are clung to longer than are hypotheses. This point is indeed reflected in the very meaning of the term *hypothesis*. We all know what it is to be hypothetical about something: Something is hypothetical if it is merely proposed as possible; that is, a hypothesis is very "iffy." A hypothesis (like the related term "thesis") is something *to be proved*; it is only tentatively held in order that it may be checked out more fully.

"Fundamental" Means "Chosen Because It Is Necessary"

A recent account of the concept of scientific "law" seems to impinge on my analysis here.[7] Scientific laws have always been sticky subjects for theorists of science. Laws have at least two features—or, better, we *behave* as though they had these two features—which distinguish them from hypotheses.

Laws are taken to be both "subjunctively forceful" and necessary. These two ideas are intimately related. Let me take them, however, one at a time. The idea of "subjunctive force" might seem like an odd one, but you already have some sort of a clue about it from our discussion of the conditional in Chapter 2. As I noted there, the conditional seems to have swallowed up some parts of the English subjunctive mood. Subjunctive mood includes what are called *counterfactuals*. A counterfactual statement is one which relates things which are not true in an assertive sense, but are still true in an odd subjunctive sense (which I cannot hope to adequately analyze for you; just try to see how the examples work). Let me give you an example. Suppose that it is a true generalization that "All the coins in my pocket are dimes"; i.e., "If *x* is a coin in my pocket, then *x* is a dime." This generalization has the logical form of a scientific law (or principle, etc.). But is it a scientific law? This question in part can be answered by determining whether the generalization has subjunctive force, that is, whether it can be translated into a counterfactual and still be true. One such counterfactual would be "If this half-dollar were in my pocket, then it would be a dime." Clearly this statement is false. If this half-dollar were in my pocket, it would be a half-dollar, not a dime. Simply placing it in my pocket would not mean that it would become a dime. The generalization thus fails to support a counterfactual based on it; we must conclude that it does not possess subjunctive force.

Also involved in this example is the idea of the "necessity" of a scientific law. It might very well have been sheer accident that the generalization "If *x* is a coin in my pocket, then *x* is a dime" is true; for

[7]Nicholas Rescher, *Conceptual Idealism* (Oxford: Basil Blackwell, 1973). See especially chaps. IV and V.

example, I might have just spent my last two quarters on a beer. But surely we do not believe that any given statement is a scientific law just because of an accident, just because things happened to work out that way. To see this, consider another example statement. Take the generalization "If *x* is made out of iron, then *x* will be attracted by a magnet." Suppose I show you one of the dimes from my pocket, and assert the counterfactual "If this dime were made out of iron (which it is not), it would be attracted by a magnet." Is this counterfactual true? Of course it is: if the dime *were* made out of iron, it *would* be attracted by a magnet. I realize that all this seems very peculiar, but there are some very good reasons that the statements work this way. Let us go into them by asking, "What do we mean when we say that a law is necessary, other than that it is nonaccidental?" There seem to me to be a couple of elements involved. First, we have a clear idea how and why the law is true. Second, we have decided that, because of its role in the grand scheme of things, the law is absolutely required to be true. These two points need further discussion.

We never assert any generalization unless we have some sort of evidence for it. In fact the logical form itself of a generalization involves some sort of correlation—conjunction—of data points, since the conditionals are manufactured from sets of conjunctions. But can this be sufficient? Clearly not. We can never be sure that we have enough data points to absolutely guarantee the jump from correlation to universal conditional. In terms of data alone, we can never make enough observations to provide a certain foundation for the law. To see this, consider a couple of examples. We believe that "Iron is attracted by magnets" is a law of wide scope—it holds for the universe. That is, it is necessarily true, anywhere, anytime, in the universe, that "If *x* is made out of iron, then it is attracted by a magnet." But our observations of iron and magnets are not universe-wide; we have been observing them only for a couple of thousand years, in a very restricted region of space. Given these severe limitations in data, why do we make the apparently reckless jump into calling $(x)(\mathrm{I}x \rightarrow \mathrm{M}x)$ a necessary law? The evidence is essential, but not sufficient, to make the jump. Thus, the evidence alone cannot warrant the jump. But the evidence is helped quiet a bit by something else: We not only have a clear idea of the generalization and its evidence, we also have at least a plausible conception of a physical structure or mechanism which might be causally sufficient for the truth of the "if, then" generalization.[8] (By "plausible" I mean that the conceived structure fits in with all our notions of other physical mechanisms—a topic which will be discussed below at more length). This conceived structure backs up the "if, then" evidence just enough to justify our belief that the conditional is a law.

[8]At this point my account deviates from Rescher's: He does not at all talk about the role of the hypothesized mechanism in providing evidence for the decision to accord law status to propositions. Harré's ideas are here at work in my account.

The Pasteur case provides a useful example. Both Pasteur and Liebig knew the conditional "If yeast is added to sugar water, then alcohol and CO_2 are produced." Liebig thought that this conditional was merely a generalization about an accidental happening, a correlation with no necessary connection between antecedent and consequent. That is, Liebig thought the generalization was just like "If *x* is in my pocket, then *x* is a dime." Pasteur, on the other hand, felt that the conditional reported something lawlike, something regular and orderly in nature. And he even more strongly felt this when he could come up with a plausible causal structure, a conceived mechanism such as the one described by the statement "Yeasts eat sugar and excrete alochol and CO_2." The general role of the plausible causal mechanism, the "underlying structure" as I call it, is perhaps best seen when we view it negatively. Suppose that we are investigating a particular conditional $(x)(Fx \rightarrow Gx)$. Suppose further that we have concluded, "There is absolutely no possible, no plausible, no conceivable physical structure which could explain this conditional." What are we forced to do in this case? Do we accept the conditional as a law, as a description of a necessary, universal regularity? I do not think so. In fact, what we probably would do is to reject the conditional itself, and start over. We would decide that we were somehow conceiving the universe wrong, even in its basic observational states such as Fa and Ga. This is probably what happened when Lavoisier considered the "law" which said "If *x* gains weight in combustion, then *x* has lost phlogiston." He could not conceive of a mechanism which involved "negative weight"; such a structure was just too implausible, it did not cohere with the concepts of the other physical structures in nature that he believed in.

Given this sort of analysis, we can conclude at this point that at least two elements are so far involved in making a law "necessary": evidence for the correlations underlying the conditional, and the existence of a plausible candidate to function as the structure which is responsible for the happenings related by the conditional. But this is not yet enough. To call a generalization "necessary" requires at least one more element: a decision which we ourselves make. That is, a generalization does not become a law unless and until we have decided that it is necessary that it be so. I am quite sure that this sounds a bit crazy. Fortunately for me, it is not quite so crazy as it sounds. In fact this conclusion is absolutely required by my account of scientific theory and practice. In part, the fact that a law's necessity comes about because of our decision was already revealed when I described what "fundamental" means in terms of the resistance scientists offer to the declaration that something in their belief system is false. That is, the more necessary the generalization, the more resistance there is to throwing it out. But there is more to our decision to make something a law than is revealed by the psychology of the situation.

Involved in this is a primitive epistemological/metaphysical aspect which must be dug out from the welter of elements. And that is: Where else could the necessity of a law come from if it did not come from our deciding that it were so? Laws, whatever they are, do not have the metaphysical property "being necessary." They do have other properties, e.g., they have properties such as "being conditional," "being of great length," "being stated in English (or German, or whatever)," and "being universally quantified." But they themselves are not "necessary" in and of themselves. Moreover, the worldly events they relate are not in themselves "necessary." The events are of themselves, e.g., what we call "sugary," "fermentative," "gassy," "alcoholic," and so on, but they are not "necessary." Yet, we still must say that laws (and their described events, objects, and conditional relations) are necessary. We seem left in the lurch: If laws have "necessity" as a property, but this is not a property of the law itself, or of the described elements of the world, then whence comes the necessity?

One way out of this bind is to make a move analogous to one we made earlier. When we were asking ourselves where the notions of "space" and "time" came from, we ended up in the fix that, although objects and events might have properties such as "red" and "loud," they did not have properties such as "before," "after," and "to the left of." At that time, we agreed with Kant that these latter properties were somehow features of conceptual systems, that is, that they were features somehow brought to the world by our conceiving minds as mental sets, predispositions, ultimate beliefs about how to conceive the world. I think that we must make a similar move here. Generalizations come to be laws when we confer upon them the status of "being necessary." That is, we *decide* to conceive them this way. To say this is not, however, to say that the world is re-created in some new way at this time just because we have decided that it is the way the law describes. Not at all. Rather, the world comes to be *conceived* to be some way, because we have decided that that is the way it ought to be. Another way to describe this is to say that the newly conferred law takes its place in the whole of our conceptual system as a functional and significant part. The new law comes to cohere within the *whole* structure as a functioning element. Note how well this view accords with what I said just a few paragraphs back. The generalization *does* have evidence in its favor; it is not simply a whim. But the evidence is not sufficient in its own to warrant our calling the generalization a "law." However, in addition to the evidence, the generalization also has associated with it a plausible causal mechanism, that is, a physical structure which we can conceive to fit in well with other physical structures that we believe in. At this point, we can decide that the generalization is, in fact, a law. Thus, it becomes at this moment necessary. We can now fit it into the whole of our conceptual structure, a fact presaged already by the plausibility of the hypothesized causal mechanism. That it is a law means that we will now behave toward it as a law, using it more or less as necessary, depending upon its place in the overall structure of our conceptual system.

This account does much to clarify what was going on in the neutrino case. There, it looked as if Pauli and his fellow scientists were just being arbitrary and whimsical when they decided to tentatively accept the neutrino hypothesis even though it violated one of the necessary methodological laws, the rule that "If *x* is scientifically acceptable, then it must meet the observability requirement." But as we have seen, Pauli and his colleagues relaxed this part of their conceptual system in order to preserve another, namely, the principle that "If *x* is a physical interaction, then *x* is conservative of energy, mass, etc." This relaxation is a drastic price to pay for a drastic move. But it is *not* a whimsical or unjustified move. Human beings themselves ultimately decide in great part how to construct their conceptual systems. And these conceptual systems are evaluated in significant ways according to how they work. When faced with a conceptual system which is in imminent danger of failing to function, it is quite necessary that its creators modify it. How they do so is *almost* entirely up to them, as a community of scientists. This is what went on in the neutrino case, when Pauli and his colleagues decided that the conservation principle was more necessary than the observability law.

CONSERVATION LAWS: THE MOST FUNDAMENTAL CONCEPTS OF ALL

I would like to conclude the present discussion by turning to a topic which has always fascinated me—the conservation principle(s). As I noted in the first chapter, the logical status of these principles is an odd one. They are taken to be of a very fundamental status (the Pauli case well shows this), and indeed, one of the foundation stones of empirical science. But they are impossible of logically valid empirical proof, since each attempt to prove them by experiment is doomed to commit some logical fallacy or another. Yet we cling to them, as if they revealed some basic primitive truth about the universe. Clearly, our behavior in this matter needs investigation.

Meyerson's Analysis: A Priori Versus A Posteriori Proofs

Probably the single best extended study of the conservation principles is that carried out by the great French historian and philosopher of science, Emile Meyerson. Meyerson's monumental research was published in his book *Identity and Reality*, which was translated into English in 1912, and immediately established Meyerson's scholarly reputation.[9] Meyerson's

[9]Emile Meyerson, *Identity and Reality* (New York: Dover Publications, Inc., 1962), p. 421.

methodology was quite straightforward. He first collected and collated together all the various threads of the conservation concepts, in the effort producing an historical survey which extended as far back and away as Hindu thought in the sixth century B.C. He then subjected his historical results to philosophical analysis. His historical conclusion is unquestionable: Threaded into all human thought, extending from the prephilosophical through the philosophical and into the scientific, the historian must find the notion that observable physical changes mask an underlying unchanging reality. This concept is found in the mainstream Greek notion which posits the view that even violent physical change, fire for example, does not destroy an unchanging identity of the fundamental physical stuff, the atoms of the substances which have burned. In this kind of case, the identities of the substances are conserved. Similarly, the medieval view that "The totality of the effect is equal to the totality of the cause" expresses the notion of an underlying identity in the interaction between cause and effect. In modern scientific times these two ideas branched out into their separate domains, namely, the view that the substances involved in change are always somehow *there*, throughout the change; and the associated view that our conceptual descriptions of these changes, our chemical and physical formulae, must always preserve a formal, mathematical identity between input and output states of the reaction. This is Meyerson's historical conclusion.

Associated with the historical conclusion is his philosophical analysis. He sees the conservation principles in science as being the explicit manifestation of this age-old human desire to find identity—of both substance and thought—and to conceive of a fundamental stability throughout change. One of the most interesting aspects of his analysis concerns the logical and epistemological status of the principles. He examines the various proofs of the principles which have been offered in modern times, and finds them all lacking. He begins by making the usual epistemological distinction between a statement which is *a priori* true and a statement which is *a posteriori* true. The a priori truth is one whose truth does not depend upon any particular empirical experience for verification. "Either there are ten neutrinos in my typewriter, or there are not" is a typical example.[10] We can determine the truth of this statement *prior* to, without need of, any examination of my typewriter. But an a posteriori truth is different from this. Consider "There are ten neutrinos in my typewriter." Clearly I cannot decide whether this statement is true until after, *posterior* to, an actual investigation into the deepest, most hidden interstices of my Smith-Corona. This example should point out fairly clearly what kind of difference Meyerson is looking into.

[10]Copi, op. cit., p. 278–279.

Meyerson approached the statement "Energy is conserved in chemical and mechanical interactions" in an effort to see how scientists argued for its truth, to see whether they appealed solely to logic and concepts, as would be the case in an a priori investigation, or whether on the other hand they appealed instead to actual empirical investigation as the basis for affirming the statement's truth. What he found was most peculiar. During the seventeenth and eighteenth centuries, great scientists like Bernoulli, Descartes, and Leibniz argued for the conservation principles on purely a priori grounds. Leibniz's argument is typical: If God is the creator, and God is also a skilled workman, then he will create a universe which will not run down, one which will not have to be fiddled with to keep it charged up.[11] But if anything were to be lost in reactions, then the universe would fast run out of power, and God would have to wind it up again. Thus, it follows that God made a perfect universe, one which conserved "fuel" and thus would never need to be wound up once it was put into motion. This argument was used to justify the principle of the conservation of *vis viva*, which matured quickly into the principle of the conservation of energy. And most significantly, mechanical equations were developed and justified as strict logical implications from the concept of energy conservation. Thus we find that, in the very beginnings of modern science, a priori arguments for the conservation principles form the logical basis for the fundamental mathematical methodology of science.

But this style of argument changes radically in the late eighteenth and nineteenth centuries. When the physicist Joule attempted to argue that heat energy and mechanical energy were equivalent, that is, that the quantities were conserved during transformation from the one type to another, he auspiciously presented empirical data to back up his contention. Unfortunately, his data were not less than 50 percent wrong, and most significantly, everyone *knew* that they were scandalously in error.[12] Yet, most astoundingly, Joule's claim was immediately accepted, with a rapidity closely matching the grudging acceptance of Pauli's hypothesis in similar circumstances a century later. Moreover, there grew a very strong movement among all Western scientists, including men as disparate as Faraday, Helmholtz, and Tyndall, which looked to the ultimate conservation and interchangeability of all types of "forces"—electrical, mechanical, heat, chemical, and even, according to the view of some, mental, ethical, and social "forces." This movement peaked around the 1860s, after it had invaded America and provided some of the metaphysical

[11]G. W. Leibniz, "A Brief Demonstration . . . etc.," in L. L. Loemker (ed.), *G. W. Leibniz: Philosphical Papers and Letters*, 2d ed. (Dordrecht: D. Reidel Publishing Co., 1969), p. 296.

[12]Meyerson, op. cit., pp. 194–195.

underpinnings for certain doctrines of social Darwinism.[13] Scrutiny of the research and writing of this period reveals clearly that the conservation principles both were leading investigators into various nooks and crannies of the experimental realm, and also were being sought as empirical conclusions of other experimental programs.

As he confronted all this diverse data, Meyerson came rapidly to see that the conservation principles could not be counted as either a priori truths or a posteriori truths. This in spite of the two fashions of proof which had been attempted during the preceding three centuries. Meyerson found himself forced to conclude that the a priori and a posteriori arguments both failed; indeed, each argument type was merely an attempt to justify a belief that was already firmly held. Since the conservation laws fitted neither category, Meyerson called their status "plausible," instead of either a priori or a posteriori. He declared that these forms of principles found the human cognition ready and ripe for belief, and thus ready to be seduced into affirming their truth. In a straightforward kind of way, he concluded, the conservation principles represent a projection of the human conceptual machinery onto the world, this projection based upon a fundamental human desire for stability and comprehension. Thus, our minds find stability and permanence where our senses find only change and instability.

Meyerson's conclusions were, of course, controversial. And they still are. However, the main points of his historical and philosophical analysis are unarguable: Western intellect, both prescientific and postscientific, has always looked for—and found—conservative systems in the physical world; but speaking in a strictly logical sense, the principles which are discovered cannot be said to be either logically or empirically justified. What, then, are we to conclude about these principles? One thing *is* clear: They form an extremely fundamental part of the ultimate paradigm of science-in-general.[14] The neutrino case gives good evidence of that. However, there is one modern addition to the conservation concept, an addition which could not have been known by Meyerson.

Symmetry Arguments: Aesthetics

In recent times, the conservation concepts have become intimately linked to what are called *symmetries*. The symmetries are conceived to be ultimate properties of physical processes. For example, it is believed that

[13]John W. Draper, *History of the Intellectual Development of Europe*, vol. 1 (New York: Harper & Brothers, 1876). The "Introduction" is quite interesting as an example.

[14]By "science-in-general" I mean the basic concept structure which I defined in Chapter 3, namely, the naturalistic system of explanation which has two goals, explanation and prediction, etc. I am assuming that there *is* such a paradigm, and, moreover, that it got its start sometime in the middle to late fifteenth century.

nature does not (indeed *ought*[15] not to) distinguish between right-hand and left-hand features of space. Further, nature does not distinguish between time past and time future; nature does not distinguish between matter and antimatter; and so on. Thus, in each case of these and other inverse pairs, nature treats them symmetrically, equally, with no discrimination for or against either member of each pair. Wherever there is a symmetry, there is an associated conservation principle, and vice versa.

It is absolutely clear that there is a strong current of the metaphysical and the aesthetic coursing through this contemporary view. If anything, the conservation principles have become more fundamental, and more explicit, than ever they were in Meyerson's analysis. They have become the ultimate presuppositional concepts in the scientific paradigm.[16] And there lies the clue to understanding them from our point of view as theorists of science. I argued above that laws gain their necessity not strictly from the correlational evidence and not strictly from the existence of a plausible mechanism, but rather from these factors *plus* our decision to treat them as necessary. Moreover, on the basis of this decision, laws come to play a coherent and highly interrelated role in our conceptual systems *as a whole.* It seems to me that we can reach a similar conclusion in regard to the conservation laws. The evidence, the experimental data, cannot strictly imply the necessity and, even more importantly, the absolute universality, of the principles. Nor, on the other hand, can strictly logical and conceptual arguments justify them a priori. But the *evidence* suggests them, *symmetry* suggests them, and we, by an act we apparently cannot avoid, *decide* that they are universally and necessarily operative. Finally, concurrent with our decision, they come to play such a fundamental role, they figure as such an integral part in our scientific conceptual system, that the system would be incoherent, and indeed would logically disintegrate, without them.

In a strange sense, the argument "Conservation principles work, and thus they are true" is right; the principles do, after all, work; science exists. But this argument dangerously oversimplifies, because it overlooks the fundamental truth that they work because we have *chosen* to make them work. Luckily enough, Nature, whatever she might really be like, appears willing enough to cooperate with our deciding to conceive of her in this way. This brings home again the fact that scientific knowledge is a uniquely human creation; scientific knowledge is not a brute creation of Nature as is a flower, a volcano, or even a human brain. But even if

[15]"Ought" is a value term, usually associated with ethical and moral contexts. But this "ought," I am almost positive, is an *aesthetic* use of the term, not an ethical or moral use.

[16]"The keystone of physics is the law of conservation of energy." So says Philip Morrison in his absolutely delightful discussion "The Neutrino," *Scientific American,* January 1966, p. 58.

scientific knowledge is a human creation, and we thus can create it the best way we know how, we are fortunate that Nature, whatever else she may be, is cooperative and, indeed, pliable in this process. And fascinatingly enough, Nature, as we have seen time and again, is resilient, much more so than our conceptual systems. Thus, for example, Nature was pliable for a time in the case of the phlogiston concept; but Nature eventually came to the limit of her flexibility. This resistance destroyed the work of Priestley and his colleagues, as the new system of Lavoisier came to be projected onto the world.

SUGGESTIONS FOR FURTHER READING

A nice overview of the "unified field" question can be found in Steve Aaronson, "Shadow and Substance," *The Sciences*, June 1974, p. 77. The whole question of "scientific law" is horribly complicated. A classical account is in Ernest Nagel, *Structure of Science* (New York: Harcourt, Brace, & World, Inc., 1961), especially chaps. 3 and 4. Stephen Toulmin, in *The Philosophy of Science* (New York: Harper & Row, 1960), chap. III, has some sensible things to say. Rom Harré's account is quite different; see *The Principles of Scientific Thinking* (Chicago: University of Chicago Press, 1970), chap. 4. I have found some of Frederick Suppe's work very useful. See his summary of the semantic view of theories in *The Structure of Scientific Theories* (Urbana: University of Illinois Press, 1974), p. 221. As far as the conservation laws are concerned, Meyerson's work remains very significant. More recent accounts would include: Gerald Feinberg and Maurice Goldhaber, "The Conservation Laws of Physics," *Scientific American*, October 1963, p. 36; and Yehuda Elkana, *The Discovery of the Conservation of Energy* (Cambridge: Harvard University Press, 1974). I particularly enjoyed this book and its look at the relation between philosophy and physics.

Part Three

Acceptance of Discoveries

The announcement of a scientific discovery is not the end of the story. To give an extreme example, Pauli discovered the neutrino nearly fifty years ago, but there is an appreciable minority of scientists who still do not believe that neutrinos exist anywhere but in other physicists' imaginations. As you can tell from this extreme example, the word "discovery" can be used rather oddly in science. How is it that a fifty-year-old scientific discovery can be still unaccepted by some scientists? In Part Three, I focus upon the question of acceptance, since it is the structure of acceptance—scientific "proof" in ordinary talk—which is at question in the Pauli case.

Acceptance, unlike discovery, has always been viewed as fair game for philosophical analysis. A wide range of opinion has consequently developed, and Chapter 8 represents my attempt to sort out these various notions. It has been agreed by all scientists that acceptance of scientific ideas must not be a strictly whimsical or idiosyncratic, subjective process. That is, an acceptance must be accompanied by the *reasons* for the acceptance. Some recent philosophers have argued that there really are no reasons, or at least, that the real reasons are flimsy and bad. But their claims have been vigorously opposed. However, even apart from this negative appraisal of the reasons, there is still a vast range of types of reasons which scientists can give for their acceptance of new discoveries. For example, some discoveries have been accepted because they represented a new and beautiful or elegant or vastly more simple approach to a problem. Other scientists have argued that these latter reasons, which are actually *aesthetic* reasons, are not good ones. Rather, they have argued, a scientist must depend on the correctness of the predictions which the new discovery allows. Yet other scientists disagree with this latter proposal; what *really* counts, they argue, is the political clout of the scientist who proposes the new discovery. As you can see, the discussion of Chapter 8 ranges far and wide. I think you will find it pretty interesting.

Following the survey, I return to our scientists, Lavoisier, Pasteur, and Pauli. Each of these men faced an uphill battle trying to get his new ideas accepted. In each case, however, I propose that specific aspects of the particular argument played the major role in getting the ideas accepted. What I claim is that each discovery was the discovery of a new object, a new kind of thing, which could be conceived to exist among the other things of the world. However, *demonstrating* this new thing was the tricky question in each case. In the Pauli neutrino case, again to take an extreme example, the demonstration attempt involved a laboratory built miles below ground, in a South African diamond mine.

But there is yet another question, a question beyond the "thingness" of the new discovery. Of equal importance—perhaps sometimes of even *greater* importance—is the question of how well the new discovery's *concept* fits in with all the other scientific concepts of the time: Does the new concept produce a harder, leaner, trimmer, more coherent set of concepts? Where the answer to this last question is "yes," I claim that the acceptance of the discovery is almost guaranteed. Chapters 9, 10, and 11 illustrate my reasons for this claim.

Finally comes my conclusion. In Chapter 12 I wrap up the acceptance question, and go back through a quick review of some earlier ideas. One thing comes clear, and that is that acceptance is most likely very closely related to the two goals of science which were discovered in the analysis of the definition of science in Chapter 3. Thus, in my conclusion, I end up saying that if the new discovery aids in the scientific goal of practical accomplishment, and if it also aids in the scientific goal of satisfying curiosity, then that discovery has a good start toward acceptance. I like this conclusion because it makes an awful lot of sense. See what you think.

Chapter 8

Philosophical Approaches to Acceptance

INTRODUCTION

Most of the efforts of philosophers and logicians to analyze science have been concentrated upon the second phase of scientific work, the phase I call the *acceptance phase.* (Other authors have called roughly the same phase "verification," or "confirmation," or "justification," or some such synonym.) There are several reasons for the heavy concentration. In the first place, it is common wisdom that discovery itself is very difficult—some philosophers say that it is impossible—to understand rationally. Second, the acceptance phase is believed to be much easier to understand, which of course draws researchers. In particular, many researchers have believed that there is a logic to the acceptance of discoveries and, most importantly, that this logic can be modeled in a useful fashion upon the logical system included in Chapter 2. While I tend to agree that the second stage of scientific effort is perhaps a little easier to account for rationally than is the discovery phase, it is still a very complicated business, and this for reasons somewhat different than those offered by many of my colleagues.

There is a third reason why the acceptance phase has been such a strong focus of scientific study, at least in this century. Historians, philosophers, and logicians have long known that human cultures often do not function in the most reasonable fashion with respect to what they believe to be true knowledge. Religious belief, for example, has often been coupled to conceptual authoritarianism and dictatorship. Other systems of concepts—e.g., philosophical dogma such as Marxism—have also functioned in an intellectual totalitarianism. But one human culture has held out the hope of opposing these sorts of historical currents. This culture, obviously, is the scientific one. Many thinkers have believed that, if *science* does not exhibit rationality in its dealings with its own conceptual systems, then *no* human culture can be expected to exhibit rationality in dealing with its own conceptual systems. To these thinkers the stakes are extremely high; they involve the choice between ultimate optimism and ultimate pessimism about the human condition. For my own part, I am not so convinced about their requirement that a qualitative distinction exists between the scientific culture, and the other sorts of cultures we humans have invented. I do, however, agree with them that the stakes are high, since if science cannot by its very structure establish an openness to new candidates for intellectual acceptance and belief, it is ultimately doomed to self-destruct from an internal conceptual self-contradiction. I will make this point more clearly (I hope) later on in the chapter.

What I would like to do is first to briefly lay out for you the incredible range of opinion about the logic of hypothesis acceptance. Then, following this wide-ranging discussion, I will return to separate discussions of our three case studies, in Chapters 9, 10, and 11, and will attempt to describe them in such a way that they will exhibit their features for us to compare with the various proposed "logics of confirmation." In a final wrap-up in Chapter 12, I will propose my own view, which you will see to be eclectic and pluralistic. It will also show, of course, strong connections to the current "established views" in the theory of science.

THE LOGICIANS

Because of the strong current of logical positivism in modern studies of science, many of the currently most popular accounts of hypothesis confirmation have a hefty logical core. These theories attempt to model scientific verification upon the permitted logical inferences of the propositional and quantifier calculus. The belief is that, on the one hand, these sorts of models are somewhat realistic views of actual scientific procedures; or on the other hand, if they are not historically realistic, then at least they are ideals of safe and secure reasoning for science to aim at.

There are some thinkers of course who believe these models to be *both* realistic and ideally safe. But at the outset I would like to make note of one point. One unfortunate problem with basing the acceptance model upon logic is that when one models scientific procedures upon the logical system, one also necessarily incorporates into the procedure any deficiencies of the logical system itself. This problem will become quite evident to you in just a moment.

The group of theorists I call the "logicians" may be comfortably divided into three rough clumps. I call these clumps the "predictors," the "falsifiers," and the "growers." [I am sure that espousers of each of these disparate views will object to *my* particular clumping; but I am also sure that they would object (perhaps justly) to *any* particular clumping; hence I see their objections to *my* clumping as no special reason to cease and desist my own clumping efforts.[1]]

The Predictors

The position of the *predictors* seems to be an eminently reasonable one. This view holds that there is a direct relationship between the degree of successful prediction of a hypothesis and the degree of acceptance that it is accorded.[2] Put simply, if a hypothesis predicts well then it becomes well established. I can even oversimplify this view (not because I *believe* that the oversimplification captures the whole truth, but rather because it seems to nicely describe the fundamental intuition which underlies the predictionist view) in the statement "If a hypothesis predicts correctly, then the hypothesis is correct." This seems straightforwardly plausible. Let me try to provide a clear, clean example of how it works.

The first step is the discovery step, namely, coming up with a suitable universal generalization of the form $x\,(Fx \rightarrow Gx)$. This conditional is discovered, let us suppose, in the typical inductive way I have described earlier, that is, by conjoining correlations and then generalizing them. The conditional might report something like "Whenever the sun is at $S_1 t_1$, and the moon at $s_2 t_2$, then there will be an eclipse within twenty-four hours" or "Whenever I double the electric current, the rat doubles his speed down the runway." Once this conditional is postulated, orthodox predictionists believe that it becomes a suitable candidate for verification, confirmation, or as I call it, acceptance. The procedure for this is straightforward: One sets up the antecedent condition F and waits to see if the consequent condition G occurs. If G occurs, then the conditional $(x)(Fx \rightarrow Gx)$ has

[1]However, as one of my colleagues has pointed out, it might be that my particular clumping is "just awful," and that's why it is objectionable. He himself, however, did not find it "awful," I am pleased to report.

[2]A standard discussion of predictive power is put forward in Irving M. Copi, *Introduction to Logic* (New York: Macmillan Co., 1972), p. 433.

one piece of positive evidence in its favor. Every time, then, that F is followed by G, that counts as further evidence that $(x)(Fx \rightarrow Gx)$ is not just a hypothesis, but actually is more likely a suitable candidate to be called a "law."

In schematic form the logic of this view is as shown below.
Hypothesis:

$(x)(Fx \rightarrow Gx)$ "If the current is doubled, the rat's speed is doubled."
Observational testing:

$Fa \,.\, Ga$ "The current is doubled and rat_a's speed is doubled."
$Fb \,.\, Gb$ "The current is doubled and rat_b's speed is doubled."

.

.

.

$Fn \,.\, Gn$ "The current is doubled and rat_n's speed is doubled."

After some agreeably large number of positive tests, the hypothesis can be considered to be established, i.e., it becomes accepted:
Accepted hypothesis:

$(x)(Fx \rightarrow Gx)$ "Whenever the current is doubled, the rat's speed is doubled."

This, in a nutshell, is the logic of the predictionist position. It is simple, clear, and logically impeccable. That is, it certainly is safe and valid from a logical point of view. But it has one drastic problem: It hardly applies to science, at least any part of science more complicated than its very first beginnings. Science, as I have argued, consists of two separate, but not independent, aspects. These are, first, the predictive/empirical, and second, the explanatory/conceptual. The predictionist view applies clearly to the first, most empirical aspects of science (as you can now see, I had malice aforethought in equating the terms "empirical" and "predictive"), that is, it describes science in its most cookbook stages. Beyond that, however, it simply does not do a believable job of analysis. Let me show you why.

Almost all the really significant hypotheses in science involve entities, or objects, or states, etc., which are initially unobservable. The reason for this is obvious. If the entities involved were easily observed, then they would already have been observed by the time significant scientific gaze was first cast their way. But since they have not been already observed, it follows that some special effort will be required to observe them. To understand this, consider the following three examples of real live scientific hypotheses:

1 "Whenever something from the air combines with a combusting substance *x*, then the burning *x* will gain weight."

2 "Whenever the lactic acid bacterium 'eats' sugar water, CO_2 and lactic acid are produced."

3 "Whenever a neutrino is given off in beta decay, there is an observed energy imbalance."

Each of these hypotheses may be expressed by the conditional $(x)\ (Fx \rightarrow Gx)$. But the actual procedure which occurs in establishing this conditional is exactly backward to that seen in the acceptance of the rat hypothesis. This is because in most cases, just like the three cases above, the F condition is not observable. For example, we just plain cannot observe the "something combining" mentioned in hypothesis 1, nor can we observe the "neutrino given off" in hypothesis 3. Pasteur most certainly could not see the lactic acid bacterium "eating," and neither could anyone else during the period while hypothesis 2 was becoming accepted. In fact, the only *actual* observable in each of these three situations is the G condition. We can, in a relatively normal meaning of the term "observe," observe weight gains, CO_2 production, and energy imbalances. Given this interpretation, then, the actual logic of the predictor view must go as shown below.

Hypothesis:

$(x)(Fx \rightarrow Gx)$ "Whenever something from the air combines with a burning *x*, then *x* gains weight while burning."

Observational testing:

Ga "*a* gains weight while burning."
Gb "*b* gains weight while burning."
.
.
.
Gn "*n* gains weight while burning."

After a suitably large number of tests (although Lavoisier, of course, did it when $n = 2$—sulfur and phosphorus), a generalization is made that $(x)(Gx)$, "Everything gains weight while burning," and this is taken as compelling evidence for the truth of the final hypothesis:

Accepted hypothesis:

$(x)(Fx \rightarrow Gx)$ "Whenever something from the air combines with a burning *x*, then *x* gains weight while burning."

There is a particular logical fallacy involved here, called *affirming the consequent.* And the main problem with it, as with all logical fallacies, is that it leads to fallacious, that is, *false,* reasoning. To put it bluntly, using observation of the consequent as evidence for the truth of the conditional which contains it is simply incorrect. It is like trying to establish the truth of the antecedent in the conditional "Whenever JFK is president, a Democrat is president" by observing during President Carter's term that "A Democrat is president." A more scientific example would be something like "If Lamarck's theory of evolution is correct, then offspring will resemble their parents" becoming accepted because of repeated observations of the fact that offspring *do,* indeed, resemble their parents. To put it most simply, the observation of a true consequent in a conditional is sufficient to confirm the truth of the whole conditional—whether or not the antecedent is true! Again we are faced with the problem of line 3 of the '→' truth table (Chapter 2), the line where (F → T) = T.

The more general point of all this is that prediction based upon truth-table conditionals simply will not do as a model for valid hypothesis establishment in any case where the antecedent of the conditional refers to an entity which is unobservable. Since this is the most typical case for discoveries, the predictionist view is in deep trouble as an analysis of real science. The root of the trouble lies in the third and fourth rows of the '→' truth table, where the conditional is true even though its antecedent is false. Thus, we might say that the two conditionals "If Darwin's theory of evolution is correct, then offspring resemble their parents" and "If Larmarck's theory of evolution is correct, then offspring resemble their parents" are both true, since they both are true; but unfortunately, biologists currently believe that the antecedent of one is false, whereas that of the other is true.[3]

[3]This problem also has another version. It is sometimes well known that the hypothesis in question does, in fact, imply the observation. In this sort of case, what is in question is not the truth of the conditional itself, but rather the truth of its antecedent. For example, it was finally discovered via logical analysis that Einstein's general relativity theory implied (predicted) that starlight would bend as it neared the sun. Thus the conditional in question would be $(x)(Fx \rightarrow Gx)$, where (F) would refer to the propositions of Einstein's theory, and (G) would refer to observations of starlight bending near the sun. On this account, the conditional was known to be true, and the starlight observations were accepted to be true. But the question still remained: Is the (F) condition *true,* that is, does the theory referred to by (F) correctly describe nature? The problem here is identical to the general case discussed above, namely, the fact that line 3 of the '→' truth table describes a true conditional which has a true consequent; yet, unfortunately, it also has a false antecedent. The root problem, of course, involves the general logic of the predictionist position itself, and not any of its particular instances in, e.g., the Pasteur case, the Einstein case, and so on. This general logic may be expressed as "If hypothesis H is correct, then its prediction P will be correct" (which can be granted to be true) conjoined to the statement "P is correct." Together, these two premises would seem to argue validly for "H is correct." Obviously, however, line 3 stands in the way of this ultimate inference about the validity of the predictionist logic.

What we find here, then, is a feature of the logical system becoming incorporated into a model for scientific procedure. What results, unfortunately, is a model which is safe for relatively trivial cases, but fallacious for more significant cases. For this reason, some logicians and philosophers of science have seized upon another type of safe inference in the logical system. These theorists I call "the falsifiers."

The Falsifiers

Sir Karl Popper, who is the exemplar of the *falsifier* school of thought, seems to have reached his position through consideration of two distinct but related points.[4] First, as I noted above, the use of positive predictive evidence does not seem to be logically capable of establishing hypotheses. Thus, he thought, it might be possible to set up an alternate system using *negative* evidence to *dis*establish hypotheses. This change in logic will assure validity of reasoning, as I will show below. A second point, an epistemological one, is related to this logical point. Science has exhibited one historical feature which is plain for anyone to see: that it often changes its mind about which theories, paradigms, and general conceptual systems are accepted. The analysis of scientific traditions and revolutions I provided earlier focused upon this fact. But Popper sees a slightly different significance to this fact than I have yet mentioned. He believes that the willingness of science to change its accepted beliefs in the face of contrary evidence is an essential hallmark of the scientific enterprise. Thus, the scientific mentality is always undogmatic, not hidebound in the way that, for example, orthodox religionists might be. Popper calls this sort of a group of like-minded individuals an "open society." For Popper, to be a member of an open society is to be always willing to revise your concepts, to be always ready to disestablish the established paradigms and theories. It is clear in his view how the logic of the situation and epistemology (indeed, even the politics and ethics) of the community are related. As it turns out, the task of science is to be as creative as possible in discovering hypotheses, and then in turn to be equally creative in the design of experiments—to show, in the first place, that the hypotheses are unacceptable, and barring that, to disestablish those accepted hypotheses whenever they show signs of becoming dogma. Unlike the predictors, then, the falsifiers never seek to positively accept discoveries, but rather to negatively falsify them. Their logic has been well chosen to suit their goals, as I will now show. I will hold until later the discussion of what looks like an historical and factual deficiency in this position.

[4]See Sir Karl Popper, *Conjectures and Refutations* (New York: Harper Torchbooks, 1965), especially the essay "Science: Conjectures and Refutations," pp. 33ff.

The predictors, as we saw, relied upon direct observation of the conjunction of ($Fa.Ga$), etc., to establish the hypothesis $(x)(Fx \rightarrow Gx)$, at least in those cases where instances of F and G were both observable. But in those cases where instances of F were not observable, they were limited to direct observation of instances of G to make their case. This move, however, I identified as being fallacious, since the conditional can be true when its consequent G is true, but its antecedent F is false. Thus, the logical fallacy "affirming the consequent" is committed in this sort of case. There is, however, a different move involving the logic of the conditional which is entirely valid, and hence secure from fallacy. This move you have already seen, in Chapter 2, where I pointed it out as MT (*modus tollens*). *Modus tollens* essentially uses the nonoccurrence of the consequent to argue for the falsity of the antecedent of the conditional. Let me show you a couple of examples in which this move is used.

A suitable example of the general logic would be as shown below.

Hypothesis L: "If Lamarck's theory of evolution is correct, then offspring will inherit the acquired characteristics of their parents."

Predicted observation: "Offspring will inherit the acquired characteristics of their parents."

Observation: "Offspring do not inherit the acquired characteristics of their parents."

Conclusion: "Lamarck's theory of evolution is false."

This argument, in effect, expresses the general logic of many particular predictions which might be tested in experimental laboratories. These particular cases would be detailed instances of this general logic. The following would be an example:

Prediction B from hypothesis L: "If the bobtail of a boxer dog is an acquired characteristic, then offspring of bobtailed boxers should inherit this characteristic."

Actual observation: "Over hundreds of generations, it has never been observed that bobtailed boxers produce bobtailed offspring."

Conclusion: "Lamarck's theory of evolution is not true."[5]

Many different particular instances of this sort of observation have been noted. For example, orthodox Jewish males continue to require to be circumcised, even though circumcision has been performed for 6000 or 7000 years. Shaving still is required for men who desire a hairless face, and so on. Thus these cases seem to indicate fairly conclusively that Lamarck's theory of evolution is incorrect, at least insofar as this particular tenet about inheriting acquired characteristics is concerned. The logic of this form of disproof is absolutely impeccable, since it relies upon the fact that a true conditional cannot have a true antecedent but a false consequent (corresponding to line 4 of the '→' truth table). On this basis alone, the logic which underlies the falsifier position is much more secure than that which underlies the predictor position. Unfortunately, however, it is not without its own idiosyncratic deficiencies.

The French philosopher and historian of science Pierre Duhem was the first to explicitly pick out the logical tangle involved in using MT to falsify hypotheses. Duhem's view fits in rather snugly with some of what I earlier claimed about scientific conceptual systems.[6] As Duhem asserts, and I second, scientific theories (and all other species of conceptual systems) exhibit logical relations among their conceptual elements. That is, the various notions are linked together in truth relations, such that the truth of one is required for the truth of another and so on. This can be described as "wardrobe" logic: If you replace one item in a wardrobe, then the other items must be replaced as well if you are to remain consistent in style. But another relation also enters into this problem, the one I described in Chapter 7 when I pointed out that some notions are more necessary, that is, taken to be more essential and fundamental, than others. This relation I modeled on the pyramid idea. Together, these two ideas provide for a real sting in Duhem's criticism of falsification. What he pointed out was that no hypothesis is logically independent of prior, more fundamental laws and principles. In fact, the prediction which functions

[5]The logical details of this situation are a little more complicated than I indicate here. The particular prediction B is taken to involve a true antecedent ("Bobtails of boxer dogs are an acquired characteristic") but a false consequent ("Offspring of bobtail boxers will inherit this characteristic"). Thus the prediction itself is a false conditional. But the nonoccurrence of the predicted observation, i.e., the false consequent of B, is not evidence *solely* for the falsity of B itself, but rather is taken to be evidence against the original hypothesis L. This is because the detailed prediction is a logical consequence of the general hypothesis. Since the implication expressed by the general hypothesis is taken to be true, i.e., hypothesis H is a true conditional, the falsity of H's consequent is taken to imply the falsity of its antecedent. This corresponds to line 4 of the '→' truth table.

[6]Pierre Duhem, *Aim and Structure of Physical Theory* (Princeton: Princeton University Press, 1954), pp. 14ff.

as the candidate for falsification [for example, $(x)(Fx \rightarrow Gx]$, cannot be independent of these other conditionals. To use a brutally clear example, suppose that the F and G in the hypothesis referred respectively to a new type of radioactive decay and its energy balance, such that "Whenever decay type F occurs, then energy radiation distribution G will occur." Now it is clear that the very measurements which are involved in ascertaining whether G occurs or not logically depend themselves upon the more fundamental principle "In every case of physical reaction, energy will be conserved" $[(x)(Px \rightarrow Cx)]$. That is, the experimenter could not depend upon his readings of the values of G energy, unless he first supposed that the fundamental conservation principle would remain in effect during the experiment. To put this situation in its logical schematic, it is obvious that G will occur when $(x)(Px \rightarrow Cx)$ and F are *both* true. Duhem's point thus indicates that the more fundamental principles and laws are required parts of a logical conjunction that forms the antecedent of the conditional of any hypothesis. Only the total conjunction will imply the predicted occurrence G. Expressed logically, then, the *actual* conditional is a bit more complicated than first we thought. In the case in question, it would look like this:

$$(x)([(x)(Px \rightarrow Cx) \, . \, (Fx)] \rightarrow Gx)$$

In English, we would say: "In every case, if all physical reactions conserve energy, *and* F-type decay occurs, then G-type distribution will be observed." Because of this unassailable logical point, then, the nonoccurrence of G-type energy distribution does in fact falsify its antecendent beliefs—but, since these are in a conjunction, which one is falsified? It falsifies the *whole* conjunction which is on the left side of the arrow; that is what the truth table tells us. But which element of the conjunction is the villain in the situation? Consider an example of this which we have all experienced. Let us say that "If the spark coil is functioning correctly, then the spark plugs will spark" is a candidate for the status of "law." Suppose further that when we turn over the engine, the spark plugs do not spark. According to the falsifier position, this should indicate that the spark coil is goofed up. However, when we take it to the shop, the coil tests OK. What went wrong? The logical error here is ours, and it involves a second, more fundamental "law" about electrical systems in car engines: "If the battery is in good condition, it will provide electricity to the spark coil." This conditional is presupposed necessarily by the earlier hypothesis linking the coil and the spark plugs. Thus, the actual conditional here is the complicated one "If the battery is in good condition, then, if it provides electricity to the spark coil and the coil is functioning correctly, the plugs will spark." And according to *this* conditional, the nonsparking plugs can reveal *either* a bad battery *or* a bad coil. The decision about which of these disjuncts is in fact true must rely upon independent investigations. This is the problem with the falsifier logic.

As you can see, this is precisely the logic of the situation that Pauli found himself responding to. When the predicted energy balance did not occur in beta decay, the scientist found himself faced with a dilemma: He had to give up at least one of his views, i.e., he had to give up either beta decay as it was presently theorized, or the more fundamental view that energy was universally conserved. The logic of the falsifiers alone told him that. The fatal flaw in the falsifier's position, however, is that the logic alone could not tell him *which one of the conjoined antecedent ideas must be rejected.* It could only tell him that *at least one* must be rejected.

It is blatantly clear that Pauli and his cohorts did not waste a single moment trying to extricate themselves from the logical ambiguity. They rejected their present conceptions of beta decay and held fast to the more fundamental principle of energy conservation. The logical muddle was immediately cleared up, but not by any help from the logic of the falsifier criterion itself. Insofar as the falsifier logic cannot itself specify which hypothetical element has been falsified, then it is incomplete as a criterion for the testing and rejection of theories. Again, in spite of its validity and logical security, the use of a strictly logical criterion founders because of its importation of features indigenous to the logical system itself. On this account, both prediction and falsification bring about logical ambiguity when used as models for testing and establishing theories.

The falsifier position suffers not only from a deficiency in its logical base, but also from difficulties in its epistemological superstructure. Popper, correctly I think, notices that scientific ideology is nothing else but transitory. That is, revolutions in scientific concepts seem to be a hallmark of science as an institution. But he infers from this fact an epistemological—and I must claim ethical and political—precept that science necessarily is undogmatic, that it is constantly in search of those experiments which will disestablish the establishment idealogy. Now I must agree with Popper that this view is a genuinely desirable *ideal* of scientific behavior. But if he intends his view even to approach the assertion of something which is historically accurate, I must argue that it is just plain false. Scientists in the past have practiced some of the most vicious skulduggery and chicanery in the defense of their establishment. In fact the political structure of science is guaranteed to produce a maximum of establishment conservatism. This is particularly true when one considers for example the publicational acceptance methodology underlying the various scientific journals, or the approval methodology followed by government granting agencies. Popper's view creates an essentially unstable institution, a science which is continually in intellectual upheaval, accepting, rejecting, establishing, and disestablishing in a sea of bubbling conceptual ferment. But such a science would be an administrative nightmare, an institutional impossibility—and even more significantly, a psychological occupational hazard of the highest order for the individual members of the scientific community. It is highly implausible that we would find a whole community population choosing the severe insecurity that Popper's ideal calls for.

Another related point is also telling. It is just plain false that scientists behave as though their task were to falsify what they presently believe. On this account the predictors, although they have bad logic, have good history and psychology. Scratch a scientist and you will find a person searching for the true, not for the false. Although it might be objected to what I am saying here that when one attempts to falsify, one is searching for the true, I think that the objection misses the point. The notion of truth which is involved in the falsifier position is that of "approaching truth as a limit." The falsifier apparently imagines that there is an infinite series of possible hypotheses, and that what science does is to eliminate the false ones so that, in the final stretches of time, we will be left solely with truth. Thus, according to the falsifier, science starts out from "Hypothesis$_a$ or hypothesis$_b$ or hypothesis$_c$ or . . . hypothesis$_n$ is true," then sets up an experiment which whacks off one of the disjuncts, i.e., "Hypothesis$_k$ is false," and must go on doing this ad infinitum, with only the ultimate hope that future science will be able to say "Hypothesis$_n$ is the only one left, and hence it must be true."

It is pretty clear that real scientists do not act this way. More likely, they behave just as the predictors say they do. The scientist starts from the position that "I believe that hypothesis$_b$ is probably correct, so now what can I do to prove it?" Unfortunately, of course, the logic of the situation is a muddle, as I have shown. Perhaps later we will be able to provide some other sort of a warranty for the positive efforts to prove hypotheses.

Even given its problems, however, the falsifier position contains something of priceless value in its core position. It is absolutely essential that somewhere in the scientific tradition be preserved the undying spirit of antidogmatism. The soul of scientific epistemology will die if it becomes committed to the view that knowledge is confined only to the established notions. But how to do this is the hard question. Certainly the individual scientists cannot be required to live in psychological insecurity. He cannot be required to be ready to attack each and every one of his personally cherished views. This simply is out of the question. However, it can be required of the scientific culture *as a community* that it carry out the sort of disestablishment activities that the falsifiers call for. Thus, we return to the "trial by fire" I referred to very early on. Attacking hypotheses must be one of the avocations of all scientists (although, of course, we cannot expect them to attack their *own* hypotheses). In the past this has been the case, and it most likely is this tendency which Popper and his colleagues have noticed. Unfortunately, such sallies against the views of friends and colleagues tend to produce personal tension, often pit department A against department B, or even department member A against department member B. But, as I noted above, disestablishmentarianism has occupational hazards. Apparently, however, there is less occupational hazard when one keeps the antidogmatic ideal, not against one's own ideas, but rather against those of friends, colleagues, and former friends.

The Growers

The final species of logically oriented thinkers is the group I call the *growers*. Although this might sound like a facetious title, I do not intend anything negative about it at all. Far from it. I believe that this position, like the first two, has seized upon an essential feature of scientific behavior, and that its fault stems mostly from its commitment to a particular logical criterion for evaluating establishment. The late Imre Lakatos is the most noted exemplar of this view.[7] He appears, from what I can see, to be most indebted to Popper's tradition. Lakatos agrees with Popper that science must be an open society, that it must have some degree of rationality underlying its changes in ideology, whatever logic that rationality requires. Like the two other logician positions, the growers believe that science must, if it is to count for anything, progress toward the truth; that is, scientific knowledge must *grow* during conceptual change, or else revolution and conceptual overthrow are purely stylistic, irrational efforts to put on a new intellectual wardrobe. On the earlier two views, scientific knowledge grew either by positive addition of newly proved hypotheses, or by decreasing the length of the chain of disjunct hypotheses each time a hypothesis is whacked off via falsification. But on this model, the grower model, scientific knowledge is not linked to some specific model of logical methodology; rather, the *evaluation* of a revolution is linked to a specific logical model—and it is a simple one at that. Let me give you the details as I understand them.

Lakatos starts out by making a plausible revision in the notion of a paradigm, or overarching conceptual system. He points out that various conceptual systems get established into institutions which he calls *research programs*. A "research program," not unlike one of my "tradi-

[7]I. Lakatos and A. Musgrove (eds.), *Criticism and the Growth of Knowledge* (Cambridge: Cambridge University Press, 1970); see especially Lakatos' essay, "Falsification and the Methodology of Scientific Research Programmes," pp. 91ff.

tions," comes to have a life of its own. It acquires its own specific concepts, its own specific experimental methodologies, its own favored questions and answers, and ultimately, its own culture and community as its members become teachers and thesis advisors and grant receivers and so on. Like a tradition, a research program continues over time and, as one might expect, gains power and influence through all sorts of sociological and political means, as well as through its scientific success in the jobs of explanation and control. Thus far, Lakatos' idea has great merit simply because it is a historically accurate model of what a great deal of science is and has been like. My suspicion is, indeed, that Lakatos' model will become even more accurate as the connection between big government and science continues. It is evident that the growers are correct just insofar as it is true that science *will* grow in the hands of big government; soon it will be "big science," if it is not already. But Lakatos' idea of "growth" is not this one, this one of growth in political and economic power and size. Rather, Lakatos has in mind the growth of *knowledge* as being the criterion for scientific rationality. He believes that the success of a research program is to be evaluated in terms of its logical relation to the historical programs which have preceded it. If there is not a logically estimable growth across the period encompassing "research program$_1$ $\rightarrow$ revolution $\rightarrow$ research program$_2$," then the revolution was not a rational one, and science has not *pro*gressed, it has *re*gressed.

Again, in this notion of a regressive revolution, we see a virtue in Lakatos' analysis. *Some* revolutions are failures; they simply do not lead to a better science. But others *are* improvements, they *do* represent a growth in knowledge. Insofar as Lakatos can make this distinction between revolutions, his analysis is more elegant and plausible than is Kuhn's, which, although it has much to recommend it, apparently cannot identify whether a revolution is really progressive, or merely a wardrobe change. As I earlier pointed out, revolutions and discoveries must be motivated by a belief that they represent an improvement: Scientists would be very foolish to accept a discovery or disestablish their beliefs without considering their new beliefs an improvement. If there is no improvement, then why change? Unfortunately, the hard question for Lakatos is now raised: What counts as an improvement? Since there have been revolutions which could count as regressive even though the scientists concerned believed that the new concepts were an improvement, it appears that we can be mistaken about the probability of improvement. What does Lakatos have to say about this crucial question?

Lakatos' view is eminently plausible, at least at first glance. Suppose, he says, that research program$_1$ can deal with x number of problems, and that research program$_2$ can deal with the exact same x number of problems plus y problems in addition. Clearly, then, RP_2 is an improvement over RP_1 since it can deal with the same problems plus some surplus.[8] This view seems eminently plausible, as I noted; but unfortunately it is rife with problems.

As has been noted by many—myself included—revolutions in conceptual systems or paradigms or research programs or traditions always involve some losses. That is, there are always some questions which, although they were previously kosher, become unaskable in terms of the postrevolutionary conceptual system. Thus, when Lavoisier's quantificational system became established in chemistry, it became impossible to ask qualitative questions such as "Why are metals shiny?" The fact that losses are involved does not eliminate Lakatos' numerical criterion from consideration, but it certainly does complicate matters almost beyond the limits of practicality. The computation now involves not only adding up the number of problems dealt with by the respective programs, and comparing them; but also, somehow, figuring in the losses. And that gets very messy. For example, how are we to compute the significance of the losses? If we let "benefits_1," "benefits_2," "costs_1," and "costs_2" stand for the relative numbers of questions answered and questions now unaskable in two successive research programs, we can set up two alternate ways of estimating and evaluating progress. Method 1 would be:

$$\begin{array}{l} (\text{Benefits}_2 - \text{cost}_2) \\ -\ (\text{Benefits}_1 - \text{cost}_1) \\ \hline \end{array} \qquad (1)$$

and method 2 would be:

$$\frac{\text{Cost}_1}{\text{Benefit}_1} - \frac{\text{cost}_2}{\text{benefit}_2} \qquad (2)$$

Certainly these two methods would produce different results. The notion of "growth" would not itself specify which of these methods (or other alternate ones) would be preferable. In any case, what started out as a simple quantitative method to evaluate succeeding RPs is now inextricably convoluted and enmeshed with questions about which evaluation technique is best (a qualitative question if there ever was one), or more

[8]My account is a little vague here, on purpose. It is not entirely clear what "deal with" and "problem" might mean. It has been suggested that "deal with" could be interpreted as either "predict" or "explain" (or both), and that "problem" could be replaced with "observational fact." Thus, if RP_1 (or theory_1 or tradition_1) can predict and explain $\text{fact}_a+\text{fact}_b+\text{fact}_c \ldots +\text{fact}_m$; but RP_2 can predict and explain $\text{fact}_a+\text{fact}_b+\text{fact}_c \ldots +\text{fact}_m+\text{fact}_n$, then RP_2 is an improvement just insofar as it encompasses a growth in our factual knowledge. The problems with any of these interpretations will be pointed out below.

fundamentally, whether certain sorts of losses (*sorts* of losses, not *quantities* of losses) are more significant than others, and so on.

But these objections only make Lakatos' program more difficult. They do not squelch it completely, except perhaps in a practical sense. There is, however, an objection—one indigenous to the logical system itself, of course—which apparently does provide a rock upon which Lakatos' ship must founder. His criterion of quantificational evaluation necessarily depends upon our ability to count the number of problems which can be dealt with under both new and old RPs. But if we interpret "dealt with" as either "predicted" or "explained," which are typical interpretations of Lakatos' intentions, then the problem becomes insolvable. It simply is not, and cannot be, known how many predictions and/or explanations may be logically inferred from a set of hypotheses, laws, and principles. And I must emphasize that this does not involve just the *practical* problem that we have not gotten around yet to counting the predictions; it also involves the *logical* problem that it is indeterminable how large a number of deductive consequences might conceivably be deduced from a set of premises as rich as those of any of the nontrivial scientific conceptual programs. It seems to me that, in at least one sense, there is an infinite number of observable facts which can be predicted from even the most low-level hypothesis. Consider the hypothesis "Whenever the voltage is doubled, the rat's velocity is doubled." If we might suppose that there is an infinite number of rats in the universe, then, if the hypothesis is truly true, there is an infinite number of true facts deducible from it. Although this is a trivial example, it makes the point. If this hypothesis were confronted with another, equally applicable hypothesis, both would imply an infinite number of cases. How in the world could it make any sense to say that either represented a "growth" in knowledge over the other? I realize that there are different-sized infinities, but since we cannot even count as high as the smaller ones, it seems to me that Lakatos' quantificational criterion must fail in the face of this objection.

As I have cautioned before, however, we must not let the success of these objections obscure the essentially correct points made by the growers. It is clear that if scientific knowledge does not in some sense *grow*, that is, if acceptance of a new conceptual system does not represent improvement in science, then science as a human intellectual activity becomes bereft of rationality. As Laudan has nicely shown, where Lakatos' thought fails is in the fact that improvement is always improvement *relative to some standard*, and the standard that he has picked, simple quantity of problems dealt with, is not adequate.[9] Thus, the

[9]Larry Laudan, "Two Dogmas of Methodology," *Philosophy of Science*, December 1976, pp. 585ff.

attempt by the logically oriented thinkers fails again, and again the failure occurs because of problems with the formal properties of the underlying logical system that provides the basis for the view. Yet, even though his specific program fails, we cannot say the same for Lakatos' general insight, just as we must reach a similar conclusion for Popper's specific program, and that of the predictors as well. Let us see whether or not we can find a more realistic program among the views of theorists other than those who are logically oriented.

THE METAPHYSICIANS

Metaphysics, as I introduced the term in the first chapter, involves our views about the fundamental objects or properties of the universe. Earlier, you will remember, we looked at the dispute between Leibniz and Newton over the question whether the universe was fundamentally atomic—made up of discrete parts—or field-continuous. The metaphysical view as applied to establishing hypotheses similarly involves questions about objects and properties. In its most essential formulation, this view asks the question "Does the hypothesis commit me to any particular belief about objects or properties in the universe?" Depending on the answer and the desire of the scientist in question, the hypothesis is accepted or rejected. One dominant view among many of the metaphysicians, a group whom I call the "new substantialists," is that espoused by Rom Harré.[10] Harré believes that new hypotheses get established when they have been shown to involve new objects or properties which can be fitted into consistent causal pictures of the observational world. Emile Meyerson, the French historian and philosopher of science, holds a view markedly similar in many respects to that of Harré.[11] I will discuss them both together. A second position, one which I am somewhat nervous about including in this section, will also be discussed. Thinkers in this position I call the "aestheticians." Included under this heading will be people such as P. A. M. Dirac, the British physicist. Dirac, as I will indicate, must be called an aesthetician because of certain of his views, most important of which is that some theories—especially, in his case, Schrodinger's quantum wave mechanics—are to be accepted because of their "beauty."[12] Dirac and other aestheticians appear to believe that the universe is simple, elegant, beautiful, or some other similar term, and that when our theories reflect this property, this is sufficient to establish them. The physicist/philosopher Leibniz also held to this position, as I will show.

[10]Rom Harré, *Principles of Scientific Thinking* (Chicago: University of Chicago Press, 1972).

[11]Emile Meyerson, *Identity and Reality* (New York: Dover Publications, Inc., 1962).

[12]P. A. M. Dirac, "The Evolution of the Physicist's Picture of Nature," *Scientific American*, May 1963, p. 45.

The New Substantialists

Emile Meyerson believed that one of the main features of science was its metaphysical discoveries. By this he meant that as theories got accepted, they tended to bring into the public consciousness ideas of new sorts of objects and their properties. Meyerson's philosophical position derives from the theories of Immanuel Kant, who himself believed that human cognition was categorically based, among other things, upon the concepts of objects, substances, and their causal relations. Kant believed that these fundamental concepts provided categories which allowed even the most youthful mind to classify, order, and arrange the incoming flux of sensations and percepts. We have already seen how Kant's views about the notions of space and time functioned. In his theories Kant, of course, was mainly talking about ordinary, commonsense objects like rocks, tables, chairs, and animal bodies (including those of ourselves and our pets). It is clear that Kant's view at least has an initial plausibility in terms of human evolution and survival. That is, if even the youngest human mind can perceive the world as ordered into objects—most significantly, objects such as its mother and father—that human mind is going to have a good jump on the problem of how to survive in this world. But Meyerson's view extended Kant's notions beyond ordinary human experiences and into the realm of scientific experiences. He thought that science's progress was the deployment and exploitation of the concepts of commonsense objects into new and very different realms. New theories had the ultimate role of providing descriptions of the kinds of objects that had been found in the exploration of newly discovered areas. This last statement is somewhat metaphorical, since by "the exploration of newly discovered areas" I do not literally mean something like a newly discovered region of the Congo jungle—although Meyerson himself uses this example—but rather I am referring to new regions such as those opened up by the hypotheses of Lavoisier, of Pasteur, and of Pauli. Just as the discoverers of the Congo brought back reports of a strange new beast which appeared to be a combination donkey and giraffe, so also Lavoisier brought back reports of a strange new substance which was contained in the air, Pasteur brought back reports of a strange new kind of life on a microscopic scale, and Pauli brought back reports of a strange new kind of very small, very fast, and almost "invisible" particle. Meyerson claimed that a new hypothesis would become accepted when it postulated the existence of a new kind of object, but an object which could be seen to be consistent with our ordinary notions of what objects were like, for example, an object which was localizable in space and time, unified, solid in some sense, persistent through time, and so on. Meyerson's view is that, ultimately, the human cognition must reduce the flux of perception—the never-ending sequence of ever-changing sights, sounds, feels, smells of the physical world—to a stable order of objects. Cognition, thus, has the job of comprehending order out of chaos, and it does this job via its use of the concepts of objects existing through time, stable in their causal interactions.

Harré's position is quite a bit more formally explicit than Meyerson's, although it clearly conforms to the same basic intuition about a main feature of science, namely, the role of science in postulating new features of reality. Let me give you an example of how their view applies. As even a careless survey of the past shows, when scientific discoveries and theories come into widespread use, nonscientists' everyday beliefs about what sorts of things exist in the world will change in order to match the new scientific theories. Examples of these sorts of changes in our beliefs about reality are easy to find. For example, prior to their general acceptance as good theoretical models of the behavior of light in certain experiments, "light rays" simply did not exist as objects of human thought. But following their initial postulation, as their features became more and more worked out, and successes in prediction and explanation accumulated within the science of optics, the idea of a light ray drifted slowly down from the scientific realm into ordinary speech and thought. In this sense, then, the discovery of light rays as good scientific models was essentially the discovery of a "new beast," one which soon took its place in both the scientific zoo and the ordinary zoo. It seems to be that this passage of concepts of new objects from pure scientific thought into ordinary thinking is a fairly general and routine one. It is simply undeniable that accepted scientific objects become commonsense objects as time goes on. We need only think of such things as microbes, atoms, subnuclear particles, acids, oxygen, DNA, ego, and gross national product in order to see the strength of the view put forward by Meyerson and Harré. Insofar as both Meyerson and Harré call our attention to this undeniable facet of scientific discovery, their view is a valuable one. But there are some problems with the view. Let me discuss them in some detail.

Harré believes that science starts from the point when a surprising phenomenon is observed. The "surprise" involved here is not necessarily the sort of surprise you encounter when the chair breaks under you or the bear jumps out from behind the tree. "Surprise" is perhaps not so good a word as "wonder," as in "I wonder why my tongue tingles when I touch my silver filling with this bit of tinfoil?" Once this sort of wonder is initiated, Harré believes that science moves in a routine, typical fashion. The observations of the phenomenon are precisely formulated, and thought is turned to the question "What sort of an object, with what sorts of specific qualities, could be responsible for the features I observe?"

This specific question is gotten at in a very specific fashion. We catalog our various *previous* kinds of objects in an effort to find something whose behavior is at least similar to what is going on in the situation at hand. Sometimes we have to mesh together two or more concepts of different sorts of objects in order to dream up behaviors which could be causing the phenomena we observe. This sort of reasoning, from well-known, past theories of objects to present unfamiliar situations requiring new sorts of objects and behaviors, is called "analogy" or "modeling." The new construct which we invent to explain the wondered-about observations is thus called an *analogy* or *model.* I prefer to use the latter term. Then, once our concepts about the model are relatively developed, Harré contends that we go out into the world (i.e., set up experiments in the lab) and attempt to find and bring back alive one of the new beasts postulated in the model. In Harré's terms, the model—the new hypothesis—refers to a "hypothetical mechanism" which at least has the status of a candidate for existence. If the conceived object provides a good explanation and, ultimately, offers the probability that it can be captured, then it stands a good chance of becoming an acceptable hypothesis. Later, if the object is actually observed, then the hypothesis is at that point established and the new object can be said to exist.

One interesting feature to Harré's account is that at one fell swoop it manages to accomplish both of the goals of science. In the first place, since the hypothetical mechanism was originally invented in order to be responsible for producing the wonder, the hypothesis *explains.* That is, it tells us why and how the surprising situation occurs. Then, given that it explains well, the concept of the hypothetical mechanism will motivate researchers to attempt to find it. Finally, if the new beast is found and brought back alive, then certainly the second scientific goal, prediction and control, will be satisifed. That is, it stands to reason that if we can *capture* one of the new objects, we certainly can control it well enough to deal with it effectively. Harré's position does seem to reflect this situation rather nicely.

There is also a weaker version of Harré's position. Harré can be called a "metaphysical realist" because he believes that science ultimately seeks to discover, via its hypotheses, what is *really* out there in the universe. In other words, his position implies that the list of scientific objects is a catalog of the things in reality. But Harré and the realist position, for reasons which I will explain in just a moment, have come under attack by other philosophers and scientists. These thinkers believe that the realist's requirements are too strong as far as reality and existence are concerned. Ultimately, they hold, all that is required for science is that the concepts of the hypothetical mechanisms function well in their tasks of explanation and control.[13] If the concepts do these two jobs well, it is unessential that we make the further step into conceiving of the actual, real existence of the mechanisms they describe. In other words, actual existence of its objects is not an essential part of an acceptable hypothesis. To these theorists, the conceptual models function only as convenient fictions—as parables or metaphors—but not necessarily as pictures of really existent objects. The name of these theorists is obviously appropriate: they are called the *fictionalists.* Let me describe for you some of the reasons underlying their belief.

Fictionalism One of the reasons that metaphysical realism is found problematical is that it appears to be limited in its scope of application. For example, there are fields which we would want to call "scientific" (at least somewhat) but which clearly do not involve hypotheses about objects. Economics is a good example. Now I realize that there are some problems with calling economics a "science." But it certainly is close to being one; at least it is "scientific" (a term I will rely upon your intuitions to understand). But consider one of the postulations of economics, the law of supply and demand. Is there an object, or even something remotely *like* an object, whose behavior answers to the name "supply and demand"? Probably there is not, at least not in the sense that there *is* an object whose behavior is described by the law of gravity. I am not entirely sure why I believe there is such a difference between "gravity" and "supply and demand." But in any case, precise reasons are not necessary at this point. The only insight needed here is the one which reveals that there is a whole range of sciences, and metaphysical realism is not necessarily germane to all of them. Harré himself is aware of this criticism, and seems at times to suggest that many sciences are not appropriately captured by the criteria found in the realist position. Most relevant, he believes, are the natural sciences, the sciences of nature such as physics, chemistry, biology, and medicine.

But even the natural sciences are not entirely comfortable terrain for the metaphysical realist. Physics, the most fundamental territory for realism, offers some ambushes and other dangers. I will mention only one. As exploration of the very, very small and very, very energetic regions of nature continues, the kinds of objects which are discovered become more and more unlike ordinary, garden-variety objects. Thus, subatomic particles such as quarks differ highly from such things as commonsense tables and chairs and human bodies. At least they do according to recent

[13]Ernest Nagel, *The Structure of Science* (New York: Harcourt, Brace & World, Inc., 1961), p. 129; Stephen Toulmin, *The Philosophy of Science* (New York: Harper & Row, 1960), chaps. III, IV.

theories. Obviously, this provides a difficulty for Meyerson and Harré right from the start. Both these men start out from the concept of the ordinary object of the everyday world, and use it as the foundation of our knowledge of the world. Thus science is merely the extension of our ordinary cognition. But in recent years as newly hypothesized scientific objects get smaller, faster, shorter-lived, and generally just more difficult to observe, our criteria for what counts as "real" and "observable" suffers some slippage. Indeed, we have already seen an ultimate sort of example of this problem in Pauli's little neutrino. What would it take to observe a neutrino? Is it *really* an object, relevantly similar to a table or chair or human body? These are difficult questions to answer, and the metaphysical realist must puzzle them out satisfactorily or his view is in deep trouble. One clever move, of course, would be to simply state that our present theories are wrong just insofar as they postulate objects ever more ephemeral and different from ordinary objects. I have a certain sympathy for this view, even though it is highly a priori and ultimately begs the central question. My sympathy appears here because of a certain softness in my heart toward two things, metaphysical reductionism, introduced earlier, and Harré's idea of "family continuity" in observation, also introduced earlier. Let me briefly indicate how my sympathy (*and* argument) is organized. The case goes like this: I certainly want to call the dog "real," as well as the flea on the back of the dog and the hair on the back of the flea. I probably even want to call "real" the cells of the hair of the flea and in turn the molecules I conceive to compose the cells. But the string is beginning to run out: Although I certainly want to say that the molecules of the cells of the hair of the flea are real, if they themselves (the molecules) are composed of objects called atoms, then the atoms must also be real, and this begins to make me nervous. But it is not quite over yet. We still have to do it at least one more time again, there to talk about the subatomic parts that compose the atom. Are these *also* real? On general principles I would not ever want to claim that a real object is composed of *un*real parts; consequently if atoms are real, then so are their subatomic parts. But once *this* move is made, I'm locked into believing in things which have lifetimes shorter than 10^{-13} second, and in general are odder even then neutrinos. The argument is logically vitiating in a strange sort of way. We start out requiring that each smaller zone, each composing part, be just as real as its parent system, thus leading ourselves on a search for the fundamental particle. But the search, which is fueled by our desire to find what is fundamentally real, ends up tunneling deeper and deeper into layers of things which are so entirely different from our ordinary world that it becomes apparent that we are really stretching to call them real. So the search for the real leads to the unreal. What is to be done?

The fictionalists believe that they have the answer. They assert that what we find in scientific discovery is not new objects, but new concepts. And a concept is acceptable or not solely on the basis of how well it works. Thus, even though our concepts *seem* to be about objects such as atoms, electrons, neutrinos, and other sorts of particles, they are not really names of such things. Actually, "atom" names an object in exactly the same way that "unicorn" names an object; "supply and demand" names an object exactly as "Sherlock Holmes" names another object. If this view is adopted, questions such as "Do neutrinos *really* exist?" simply do not come up. Moreover, when we decide to drop a theory, i.e., to disestablish a hypothesis, it is not as if we were killing off an old and tired family pet, but much more as if we were merely letting a poor story go out of print. On this account, then, phlogiston did not just one day cease to exist as if Lavoisier had killed it off; rather, Lavoisier merely told a more likely tale about what was going on in combustion and smelting. Lavoisier's story-telling abilities were simply better than those of Stahl and Becher.

Fictionalism squares nicely in yet another sort of problem area. In some cases, conflicting models are used in a science. For example, it is well known that "light rays" can be used to very nicely describe shadows and the behavior of lenses; on the other hand, "light waves" can be used to nicely describe the fuzziness of shadows' edges and the colors of the rainbow; and finally, "light particles" (photons) can be used to nicely explain how photocells and photosynthesis work. But there cannot be existent in the world—at least so far as we can conceive of it—some one single kind of object which combines the qualities of a ray and a wave and a particle. So whatever light *really* is, it is not a "ravicle." The problem is that all three theories provide very nice tales about various delimited regions of optics. That is, when we are concerned about lenses, we conceive the world *as if* light were *really* a ray; ditto waves when we are concerned about rainbows. But since we fully well know that light, if it exists, is some one, coherent single kind of thing, deep in our hearts we must thus really believe that each of these three conflicting light-objects is just a figment of our imagination. Thus the virtue of the three objects is that each provides a convenient fiction to explain what is going on in its respective area. The fictionalist analysis seems to fit very well here, since it allows us to apply three unmatchable sets of concepts in the same area, even while we fully well realize that the world could not contain all three. Realism, unlike fictionalism, would feel a real tension in the face of this problem, since the realist must believe that science reveals the elements of reality, and how can ray, wave, and particle *all* be real?

Problems of Fictionalism But fictionalism, even though it is a very skillfully built philosophical position, contains two completely crippling flaws. In the first place, it is not realistic as a description of real live science. Science simply does not behave as if fictionalism were true. Thus, fictionalism provides not a description, but a prescription for science. Some philosophers think that this is quite all right since they hold that the philosophy and history of science are two very different things. Something could be a good philosophical *pre*scription for science, even though it provides a poor historical *de*scription of science. However, even if we buy this view, we find ourselves in a real bind for a second reason. This is because the fictionalist prescription is fundamentally nonscientific in a philosophical sense. Consider the light case again. We all know that light *really is* something, even if that something is most definitely not a "ravicle." Light just has to be something or other; after all, we can see it. And see by it. Consequently, the nonexistence of a "ravicle" shows only that our knowledge about light is incomplete, not that light itself is a convenient fiction. Here is where fictionalism's antiscience comes in. Since we know that our knowledge is incomplete, our task must be to keep searching, to keep thinking, to keep trying to find out what light *really* is. Fictionalism, however, does not require this continuation of scientific study; fictionalism allows us to rest complete whenever we have a tale likely enough in terms of its abilities to predict and explain. Fictionalism does not require anything more than this, it does not preserve room for the motivation to press on with the search. Realism, on the other hand, with its commonsense desire to find out what is *really* going on, does provide such an unquenchable thirst. But to this, of course, the fictionalist can retort that, even if the thirst *were* quenchable, there is not much use in looking for the Holy Grail to be the cup which would hold the quenching fluid. So the battle goes on. This battle, as you can probably tell, is one which I cannot stay out of and, indeed, am obviously in the middle of already. But let me leave things as they are here, and get back into the fray—which I will definitely try to drag you into as well—when we return to our men Lavoisier, Pasteur, and Pauli.

The Aestheticians

Aesthetics is the study of "beauty" (or "perfection," "fitness," "elegance," or any other of a whole series of synonyms). At first it might seem strange that science and aesthetics would be in the least related. However, there has been a close and definite connection between them since the times of the earliest Greek science. This connection is most evident among scientists of a mathematical bent, perhaps since mathematics itself pays close attention to the so-called elegance of proofs. In mathematics, "elegant" can mean "the simplest." For example, in my own introductory logic classes I tell my students that I am not looking for elegant proofs, but merely correct proofs. But from time to time, some logic whiz will produce a proof which is both correct and extremely short and simple. At least, it is shorter and simpler than the proof which *I* can provide. This sort of proof I call elegant. Apparently, although I cannot swear by it from my own experience, elegance in the case of advanced formal mathematics is similar to the kind I see in my logic classes, but just a bit more complicated. As you can easily see from this discussion so far, it is very difficult to say much about what elegance, or simplicity—or beauty for that matter—really is. But the strange thing is, many if not most scientists know what is being talked about when someone else claims that a particular proof or theory is "beautiful."

Aesthetics, then, as applied to scientific hypotheses, involves some formal property, and this property may be called "simple" or "elegant" or "beautiful" or some other such value term. The aestheticism position on theory acceptance pushes the aesthetics/science connection even further, and this is what is important for our present discussion. According to the aestheticism view, a theory is acceptable or not acceptable on the basis of its formal aesthetic values. Thus, to put the matter in brutally frank terms, the beauty of a theory implies its truth. I am sure that this view sounds quite odd to you—not to mention that it probably sounds unscientific in the extreme. How can such an attitude be justified? Let me first say a thing or two more about it in explanation, and then give you a couple of quick examples of how aesthetically inclined scientists function.

The view that the beauty of a hypothesis implies its truth can only be justified on the basis of a specific metaphysical theory. (This is why I have included the aestheticians in this discussion of the metaphysicians.) The aestheticians believe that the truth of a theory is linked to its beauty simply because they believe that the universe itself, somewhere in its deepest, most fundamental structure, is beautiful. Thus, when our theories are beautiful, it is most likely that they are presenting an accurate picture of the beautiful reality. The view is as simple as that. Even though it sounds wild, I must admit that this position has its attractions. Let me give you a couple of examples.

Leibniz, as we saw in the first chapter, concluded that the pathway of light reflection *must* be such that the angle of incidence is identical to the angle of reflection. He reached this conclusion on the basis of his discovery that this pathway was the simplest, the most energy-conservative, in short, the *optimal* pathway. Optimality, according to his view, is a sort of physical elegance. It is the sort of feature found in the work of the best architects. His argument in favor of optimality was an interesting one. Since God created the universe, he said, why should He elect to create a universe which was *not* the best possible, which was not the most efficient, the most elegant, and the most beautiful possible? It would be irrational, he concluded, for God *not* to create the best possible universe when it was in His power to do so. No one would choose to do a sloppy job when it was just as possible to do a slick, clean job.

There is, of course, one main problem with Leibniz's argument. He requires us, before accepting it, to first accept the existence of God the creator. And that is a big request to make of modern scientists, who naturally would prefer to keep their naturalistic science separate from their supernaturalistic theology (assuming that they believe one). But still, even given this major flaw, there is a fundamental kicker to Leibniz's argument: It *does* make sense. After all, if there *were* a creator (any creator), it would make sense for Him to do the best job possible in His creating. I am not really sure what the connection is here, but I suspect that there is a link between "elegance" in a physical system and "making sense" to the human mind. One modern physicist, C.F. von Weizsäcker, a German astrophysicist, indeed has argued along Leibnizian lines that the universe must fundamentally possess the property of "intelligibility."[14] Although he does not do much of a job of describing what he means by "intelligibility" (it is not clear that he or anyone else *can* do such a job), it is clear that he believes that the universe, in order to make sense, must make sense (that is, make sense to us knowing humans). For example, it just would not make sense if the universe wasted energy at every turn in its physical interactions, if it always took the most obtuse pathways instead of the simplest and most elegant, and so on.

The curious thing about this view is that it has had a certain amount of success as a criterion for appraisal of scientific work. That is, Leibniz's view, no matter whether true or not about God, certainly turns out to be true about light reflections. They *do* follow the optimal pathway in terms of energy conservation, causal power, and so on. The analysis works for the shapes of riverbeds, certain kinds of collisions, etc. Von Weizsäcker points to further examples in which the aesthetic view is used as a heuristic principle, that is as a tool for leading research into different, and otherwise unexpected, areas. He is particularly interested in Hamiltonian functions, which are special mathematical analyses that can be related to the kinds of optimal (and intelligible) pathways mentioned earlier. He notes that, if we can apply a Hamiltonian analysis in a new area, we have good initial evidence that it is an accurate analysis, simply because it is the sort of function that it is. Thus, because the universe has its particular aesthetic feature, any conception like the Hamiltonian which matches or implies or describes this aesthetic feature is imbued with a kind of a priori acceptability.

[14]C. F. v. Weizsäcker, *The World View of Physics* (London: Routledge, Kegan Paul, Ltd., 1952), chap. VI.

P. A. M. Dirac offers the most outrageous modern example of this sort of reasoning. When Dirac first read Schrodinger's theory of quantum mechanics (the theory about the behavior of the smallest particles), he proclaimed that it must be true, simply because it was so beautiful. Dirac did this prior to *any* real experimental testing of the theory. It is obvious from what he says that he believes that the universe itself is beautiful; thus, any correct theory of the universe must reflect that beauty. Since Schrodinger's theory is beautiful, it is acceptable.

We see a heuristic use of aesthetic reasoning in many modern cases of discovery which are based upon the assumption of symmetry in Nature. Thus, some new theories in elementary particle physics have been developed on the basis that an elegant Nature would not discriminate against left- or right-handedness, or clockwise or anticlockwise spin, or future- or past-directed time processes. But such examples are not limited to physics. One of the most fascinating uses of an aesthetic heuristic is portrayed in a recent program from the "Nova" science series on TV. In this episode, the attempt is being made to explain various features of dinosaurs, on the basis of inferences made from facts about their fossils in conjunction with more general theories. One particularly elegant inference involved the role played by the series of oddly shaped fins which are found along the spine lines of many dinosaurs. A scientist working with these fins asked himself, "What could they be used for?" (Note that he *already* assumes that they *have* a use, i.e., he assumes no wasted effort on Nature's part!) Since his general hypothesis involved the notion that the dinosaurs and modern birds might be evolutionarily related, he noted that birds often raised their feathers as a means of dissipating heat. He then asked himself the following question: "If I were to design an optimum apparatus for dissipating heat from a dinosaur's back, what would the design look like?" As it turns out (it of course *has* to turn out this way, or there would be no point to my story), the fossil fins are just perfect for the job: They are staggered in just the right fashion, and their rather peculiar shape (rounded off rather than with square corners) is thermodynamically perfect to bring about maximum heat dissipation. In other words, the dinosaur fin design was an elegant engineering structure, it was an optimal design for its purpose, and most important, it made sense. Moreover, the fashion in which the question was asked made sense.

I realize that this discussion is rather fuzzy and a bit of a soggy waffle. I really do not know what to do about it, other than what I have done. It is quite clear that there is a strong—and fertile—current of aesthetic appreciation coursing through the river of scientific thought. Any realistic discussion of the means by which scientists appraise theories must mention that the appraisal is oftentimes rife with aesthetic evaluation. Moreover, it must be pointed out that the use of aesthetic evaluation presupposes a powerful metaphysical belief, namely, that the universe is such as can be described only by aesthetically valued theories. But beyond saying this, I am just not sure how much can (or ought to) be further said. It is enough, I think, for us to be quite aware of the aesthetic currents themselves, without being able to say too much specific about them in advance of studying them individually. What is more important, once we have found an instance of aesthetic appraisal, is to be especially attuned to analyzing it in its fullest and deepest extent. Only in this way can we become fully acquainted with the individual beliefs of the scientist in question. I would never say that aesthetic reasons are *bad* reasons for accepting theories; but then again, I am not sure in general why I might want to argue on the other side that aesthetic reasons are *good* reasons for accepting theories. But I suspect that they are.[15]

THE CONCEPTUALISTS

The logicians and metaphysicians are joined by their fellow philosophers, the epistemologists, in providing a view about theory acceptance. I wish here to talk about only one version of an epistemological position, the view I call *conceptualism.* My use of this term is somewhat idiosyncratic, and so I had better define it. First I need to discuss theories about what "true" means. There is a theory about truth which claims that a statement is true if it *coheres* with other (indeed, with all other) statements. That is to say, logical consistency between a new element and a preceding system of elements is a guarantee of the truth of the new element. Although this view has some obvious problems, it also has some obvious advantages. Let me describe both in just a bit of detail, since the conceptualist position depends significantly upon the coherence theory of truth.

Two Theories of Truth

The two main candidates for theories about truth are the coherence theory and the correspondence theory. The latter is the one all of us are most familiar with: A statement is true if and only if it corresponds to what it refers to. For example, the statement "Today is Wednesday" refers to two entities, namely, (1) today; and (2) a calendar object called "Wednesday." On the correspondence view, the statement is true if and only if there is a correspondence between the statement and those two entities, today and Wednesday. If there is no correspondence, i.e., if what

[15]I am not *quite* so unsure as I might seem. Since I believe that a large element of our picture of Nature is a function of our own ideas and concepts, it is clear that, if we as thinkers have an aesthetic dimension (which is unarguable), then necessarily Nature insofar as she is our conception must reflect that same dimension.

"today" names is in fact Thursday, then "Today is Wednesday" is false. Similarly, "It is raining" is true if and only if it corresponds to a situation in which water is falling from the sky, the streets are wet, and all those other things we normally perceive during the rain. Correspondence theory is relatively intuitive, and most likely is the theory we all believe up until the time philosophers start to mess around with our minds. Once they start to do this, deficiencies in our intuitive views are exposed. Let me give you a couple of easy examples. Philosophers and mathematicians talk to one another quite a bit, which may be an unfortunate fact for human intellectual history. But once these conversations got started back in the time of Pythagoras, Plato, and their fellow Greeks, it was too late to stop the process. Plato asked several good questions of his friends the mathematicians, and the world has never been the same. Consider, to paraphrase Plato, the numerical statement "2 + 2 = 4." Certainly we must all admit that it is intuitively true. But how can we explain its truth in terms of the correspondence theory? If it is true, then "2" must refer to some object, and so also must "4," "+," "=," and so on. If there *are no* such objects, then the statement cannot correspond to them—in fact it cannot correspond to anything at all—and hence it cannot be true. This follows from the fact that a statement is true only if it corresponds to what it refers to, and "2 + 2 = 4" does not refer to anything, since there is nothing there to be referred to. Thus, how can we possibly claim that "2 + 2 = 4" is true?

This problem is a staggering one, with enormous implications, most of which I will try to avoid like the plague at this time. But I must discuss it just a bit, so let me do that by pointing out two options which can provide escape from the difficulty posed by Plato. Option 1: Suppose that there really *are* objects referred to by "2," "4," and so on, but that they are very different from normal material objects such as rocks, people, tables, and suchlike. Plato himself took this escape route. He proposed that these objects really existed, but in a world quite different from the material one; it was an ideal sort of world, immaterial, eternal, unchanging, and perfect. (As you may have already suspected, such is the manner in which the initial elements of the Western ideas about heaven were launched, later to be joined and enhanced by those of St. Paul, St. Augustine, Boethius, and their friends.) Although his postulate was a bold one, Plato's new world of ideal objects *did* have the saving grace of rescuing the correspondence theory of truth from the attacks of busybody philosophers. If these objects exist, then "2 + 2 = 4" is true because it corresponds to what it refers to.

But there is another option, option 2: Simply give up the correspondence theory of truth. If this be done, then one need no longer believe in the existence of a nonmaterial, ideal world, a world peopled by "2," "4,"

"+," and later, of course, Lucifer and all the other angels and heavenly beings. Many philosophers and mathematicians have felt more comfortable doing without this other, ephemeral world. But a further difficulty remains: What theory can we put in the place of the now-discarded correspondence theory? Coherence theory seems to be the most likely candidate: "2 + 2 = 4" is true simply because it coheres, that is, it is logically consistent, with the entirety of the conceptual system of arithmetic. "2 + 2 = 5" is not true, is false, because it is not logically consistent with the rest of the arithmetical system. On this point, then, coherence theory is very useful for the philosophy of mathematics.

Later studies, especially studies into human perception and its relation to other epistemological questions, added further dimension to the uses of coherence theory. It was a reasonable move to make, following the scientific hypotheses about sensations and perceptions made by early modern thinkers like Descartes and Locke, to claim that each human being was in contact with the real world only via his own individual perceptual system. Moreover, the claim continued, it is clear that we have no guarantee at all that each of our perceptual systems is not entirely individual, that is, personal and idiosyncratic. Scientific evidence was generated to back up the claim. The existence, for example, of color-blind people points to the reality of individual differences in perception. Given this analysis, it seems clear that the correspondence theory of truth is in big trouble, since it becomes quite difficult to decide *whose* empirical world is the world which must be referred to by true statements. Certainly the perceived world of color-blind people is quite different from my own, at least in their use and understanding of color terms and concepts. But must we say that their statement "This apple is red" (which, by the way, is a sort of discrimination that even red/green color-blind persons can make easily) is false simply because we know that their perceptual correspondences are vastly different from mine? It does not seem to me that we *can* call this statement of theirs false, especially if it is true that the apple really is red. What to do? The answer is fairly straightforward: Require only that each person's statements cohere, i.e., be logically consistent, with all the rest of his or her own, and with all the rest of ours. Thus, in no case do we require, or even inquire into, the similarities or dissimilarities between each of our own individual perceptions or concepts. Such inquiry would be useless. All that is required is the logical consistency, the coherence, between the statements and behaviors of all of us. In case of clash, when someone's statement is logically inconsistent to those of others, we do not inquire what his or her perceived world is like; rather, we simply note that the statement is inconsistent with those of the others, and thus it is false.

Coherence theory does a nice job in those problem areas it was designed to solve. This is not to say that it does not have its own special problems, since it certainly does. For example, what about a mass hallucination, i.e., a situation in which everyone else is in fact wrong? In that case, the logically incongruent statement is not false, but actually is true. There is a further difficulty—an even more puzzling one—raised by the mass hallucination case. Consider the mere fact that we can even describe it for discussion.

If we can discuss it, that is, describe a situation in which the logically inconsistent statement is *true*, then it seems clear that "truth" and "logical consistency" are not exact synonyms, and thus the coherence theory is not an exact replacement for the correspondence theory. This is a serious deficiency in the coherence theory. But unfortunately we are now getting beyond the edge of the domain of our present focus, the conceptualist position on theory acceptance. Let us return to that topic.

Coherence

At any given time in a scientific tradition, there is a large body of theory, principles, laws, etc., which are well accepted by the scientific community. These ideas form the paradigm conceptual system. We have discussed this fact before, but I would now like to extend this idea of a "background of concepts" into the analysis of theory acceptance. My point is this: According to the conceptualistic view of theory acceptance any new candidate for acceptance must cohere—in any or all of several different ways—with the conceptual background, the paradigm, of its tradition. These different ways of cohering crop up in relation to the different types of discovery situations.

The simplest case of coherence is found in the case of the routine or *predicted discovery*. A predicted discovery involves the minimal sort of hypothesis. Everyone, for example, expected Goodyear to come up with a process for hardening rubber, or Edison to come up with a suitable material for the electric light. When the discovery was made, it was no big deal, simply because the facts of the discovery were in no way logically inconsistent with the body of science and engineering which had paved the way for the discoveries themselves. The function of coherence in this kind of case is easy to understand, and presents no real problem at all. The new discovery coheres simply because it is, for all practical purposes, logically predictable from the paradigm. I call this kind of coherence *predictive coherence*.

Internal Versus External Coherence

But there are exciting cases, cases which can be used to point to more profound uses of coherence between a discovery and its background. Although there are two different types of coherence—which I call "external" and "internal"—they are both described rather nicely by physicist Richard Feynman's statement in a recent TV interview, "I know that the hypothesis is a good one if it ties together and makes sense out of stuff I knew earlier but couldn't quite understand." The "tying together" that Feynman is talking about is of course coherence, and it can be of two kinds, as I mentioned. *Internal* tying together refers to the coherence within a specific field which is contributed by a new hypothesis; *external* tying together refers to the coherence which obtains between the specific field and other fields. Let me give some quick and easy examples.

When Pasteur suggested that fermentation was the observable result of the life cycle of a living bug, it made a lot of sense out of all the data related specifically to fermentation. For example, it explained and tied together the three apparently unrelated facts that: (1) adding a starting mass of yeast aided beer fermentation; (2) beer fermentations could not be started in unoxygenated, closed containers; (3) warmer temperature speeded up the reaction, but a hot temperature stopped the reaction. Pasteur's hypothesis tied together these specific facts, and gave them internal coherence. That is, it now made sense to say that the starting mass of yeast is a "colonizing population," which must breathe, and which functions better when it is warm, but which dies when it gets hot. Pasteur's hypothesis, then, not only coheres with the data of fermentation as a specific field, it makes these data *more* coherent than they were prior to the hypothesis; more, at least, in the sense that these data are now predictable from the hypothesis itself. Thus, internal coherence can be extended into "predictable" coherence.

Pasteur's hypothesis also produced a great deal of external coherence. That is, it linked together for the first time fermentation data with data from, for example, optical crystallography (remember the polarization experiments) and medicine. In medicine, for example, Pasteur's theory about bugs and fermentation could now be tied into a theory about bugs and disease. In this way the two very diverse fields of brewing and medicine could cohere together via the idea of a microorganism. What is more, the two different fields become linked together by predictive coherence on the basis of the microorganism hypothesis.

The two types of coherence figure significantly in the conceptualists' analysis of theory acceptance. They argue that a new theory is acceptable if it coheres with the background of concepts, that is, if it is either internally or externally consistent with the prevailing tradition, or both. This way of stating their view is not exactly the position of any given conceptualist, but rather is kind of an oversimplified statement of my own. But even though oversimplified, it faithfully illustrates the main problem of the conceptualist view. The problem is this: As expressed, conceptualistic theory acceptance is an inherently conservative procedure. If what is required is consistency between old ideas, between the old tried and true concepts and the new ones, then necessarily no oddball concepts will be acceptable. Such conservation, however, is not the sole deficiency. Two related problems are also raised. First, consider a weaker variant of the mass hallucination problem. Let us suppose all the present scientists are wrong about a particular theory they believe. If that were the case, then requiring that new hypotheses be logically consistent to the prevailing theory would be requiring that all new ideas be false in order to be acceptable. Certainly there is something wrong with *this* notion. Second, in addition to this odd logical problem, a historical problem is raised, namely, this version of conceptualistic acceptance has to be historically false, since it is obvious that none of the discoverers of new scientific concepts, Lavoisier, Pasteur, Einstein, etc., proposed concepts which were consistent with the prevailing conceptual background. These two problems apparently torpedo the simplistic version of conceptualism. How can we come to the aid of the view?

In the first place, I want to point out that we *should* come to the aid of the conceptualistic analysis, and this most simply because it is an accurate rendition of a very important element in the acceptance of theories. Thus, just as in the cases of the various logicians, metaphysicians, and aestheticians, there is a significant kernel of truth in the conceptualistic position; and we do not want to reject that kernel just because the position has a problem or two. Having said this, let me now attempt to rescue the conceptualistic view from mortal peril in order to preserve it, wounded but still quite alive. As in all such circumstances, my rescue will be effected by making the position more complicated.

A Realistic Conceptualism

The thing to note about scientific paradigms, about the conceptual background of a discovery, is that the concepts are structured into layers. In an earlier discussion we looked at the notion that hypotheses are more hypothetical—believed more tentatively—than laws; and principles are held to be more accepted and necessary than laws; and so on. This notion has significant bearing on the conceptualist criticism, since when one encounters the coherence question, it is always necessary to look at which levels of paradigm cohere and which levels conflict with the new discovery. Hypotheses which are accepted will almost always be found to be contrary to other hypotheses, but at the same time will mesh well with the more basic laws and principles. Because of this fact, we can easily see how it was, for example, that Lavoisier's oxygen hypothesis was not internally coherent with his chemical paradigm, and indeed contradicted it, but at the same time, exhibited an extremely deep level of coherence to the physics of his time. Thus it was that oxygen, considered as a substance with positive gravitational mass, fit very closely to the fundamental law of gravitation in physics while phlogiston with its negative weight could not be matched into the physical paradigm.

This analysis seems to straighten out the apparent paradox we were falling into, namely, that the oversimplified version of the conceptualist view required logical consistency, and thus both conservatism and no conflict with prevailing views, whereas the historical facts of discoveries clearly exhibited that both innovation and conceptual conflict were the general rule. According to our more complicated version of conceptualism, then, a hypothesis is acceptable, even when it conflicts with other accepted concepts, on the condition that it can show coherence with other, especially deeper-level, concepts. Even though this makes a decision about acceptability much more tricky to reach, given that the decision requires judgments between amount of coherence versus amount of conflict, it appears to me that such complication is required in order to save both the valuable insight of the conceptualists *and* correspondence to the historical facts.

To sum up the conceptualist position we need only note that a hypothesis is acceptable if it coheres with the prevailing paradigm. Its degree of acceptability increases directly with increase in coherence. In cases where conceptual conflict occurs between the prevailing paradigm and the new hypothesis, deeper-level coherence can make up for conflicts between higher-level concepts and the new hypothesis. Coherence itself can be of two types, internal and external. Obviously a hypothesis which produces both types of coherence is to be preferred. In cases where not both are present, then it must be said that internal coherence is *necessary* for acceptability; and although I am going a bit out on a limb in saying this, it seems that the actual behavior of scientists indicates that external coherence is most likely a *sufficient* condition of acceptability.

This summary, though complete in itself, does not completely finish our consideration of the conceptualist position. In my review of the historical cases, I will often return to the coherence theme. As you will see, growing coherence itself usually leads to growing confidence in the hypothesis in question.

THE SOCIAL SCIENTISTS

Our final discussion of the problem of theory acceptance will center on a group of thinkers whom, for lack of any better name, I call the *social scientists.* This group represents somewhat of a break from those we have looked at earlier. A main difference is the fact that the social scientists make no effort at all to *prescribe* the conditions involved in theory acceptance. Rather, they attempt only to *describe* what is going on during the period when a new hypothesis is accepted. Let me try briefly to make this distinction a little sharper.

When we investigate the logician and metaphysician positions—any of them—we note that each is directed toward finding some criterion or set of criteria which may be used to tell scientists *when they ought to change their beliefs* and accept a new theory. Thus, we can say that the logician and metaphysician are looking for a prescription which can be used to tell us what to do in the crucial situation during theory crisis. On the one hand, the predictionist believes that a theory which makes lots of accurate predictions ought to be accepted, whereas the falsifier believes that a theory which makes false predictions ought to be rejected. On the other hand, the metaphysician hopes to eventually come up with a prescription for theory acceptance. For example, the aesthetician holds that the beautiful theory is the one to accept.

The social scientist, however, makes no attempt to come up with a prescription for scientists to follow in their dilemmas about which theories to accept or reject. For the social scientist, it is more important that we get to a hard-core, correct description of the actual mechanics and dynamics involved in theory acceptances. You will probably notice this sort of a stance in the following description of social scientists.

The Politicians

Some social scientists have been impressed by the fact that, within the scientific power structure, it is "politics as usual." That is, they have studied the scientific society and found that it has exactly the same kind of institutional dynamics as does any other political institution. Thus, according to their view, getting a theory accepted is no special problem; it engenders no difficulties different from the difficulties involved in getting a policy accepted in any other sort of political economy. To give an analogy for this, then, I might say that these theorists believe that establishing the hypothesis "We ought to change our policy toward China" in the American government is relevantly similar to establishing the hypothesis "There is some substance from the air involved in combustion" among members of the chemical hierarchy.

The notion of "politics as usual" has some interesting ramifications. Moreover, these ramifications are just what we might expect from our own everyday experience in the workaday world. Let me take a simple case. As we all know, getting hired on a job is not always a matter of "how good you are" but rather is a function, often in great part, of "whom you know." If you know someone influential, then the chances are much better that you will find and be agreeably accepted in a good job. That is, knowing someone is not a *necessary* condition of getting a good job, but it certainly is taken to be a *sufficient* condition for such success. The "politician" view of theory acceptance in science is analogous to this. Suppose that you are a newly minted scientist, just graduated from a good school, working at a good laboratory, under the supervision of a very influential scientist (one who, incidentally, is the editor in chief of the most important journal in the field), and so on. Suppose further that you have created a potentially controversial new hypothesis in your field, a hypothesis which goes far beyond the routine and everyday. What are your chances of having your new theory accepted? The cynical element in each of us is immediately aroused by such circumstances. The politician's case almost makes itself in this sort of situation. But let me make the case even better for the politician. Suppose finally that there is only one competing hypothesis to yours; that is, suppose that, in regard to the observations in question, only two explanations are candidates for acceptance as the explanation of the phenomena in question—yours, and a competing hypothesis put forth by a completely unknown scientist, from a podunk school, who is working alone and does not even have a colleague, let alone a well-placed boss. Now we can ask the question again: What are *your* chances of having your new theory accepted versus the other scientist's chances? We apparently do not even need to think about it. It seems to me undeniable that all of us would a priori predict that the hypothesis offered by the well-placed, well-entrenched individual stands a better chance than does the hypothesis put forward by the unknown. On this account, then, the politicians have scored an important point for their view that it pays to know the right people and to be in the right place, as far as getting one's theory accepted is concerned. But there are problems with the politician view, which I must now mention. I will do so in the context of a fascinating story related to me by one of my good friends in the philosophy of science.

Problems with the Political Approach This good friend (a major leaguer who justifiably commands a good audience and a fair reception) had come into contact with reports of a Bulgarian's experiments which purported to show the existence of absolute space.[16] Upon first hearing of the experiments, he set himself the task of showing where the Bulgarian had gone wrong, since it was clear that he *was* wrong, that he *had to be*

[16]My friend, who shall remain nameless, at first objected to me when I pointed out that it was important that the Bulgarian be a *Bulgarian.* "It makes no difference to the point," he claimed. Of course I thought him wrong, and I so argued. As I noted, the story would not have near the sharpness that it does were the scientist to be Chinese or Indian, let alone Chilean, Australian, or German. It is important, for obvious reasons, that the scientist be Bulgarian, since that undeveloped country is well known for its isolation even from the socialist bloc of nations. There is no prejudice against Bulgarian scientists involved. I intend nothing here but a factual analysis of the perceived status of Bulgarian science, whether or not this perceived status is correct—which it might very well not be, as you shall see.

wrong. My friend thought this way, not because the Bulgarian was a Bulgarian, but because he was espousing the concept of absolute space as an explanation for the experimental results; and not since the time of Einstein has it been fashionable to refer in any way (except negatively) to the notion of absolute space. But here was this Bulgarian scientist, as reported in some obscure Eastern European physics journal, making claims that the concept of absolute space was requisite to interpret his experimental results. There is no doubt, then, how my friend came to set himself about finding where the Bulgarian had erred. Of course—or my story would not be a story at all—he could not find any mistakes in the interpretation. Indeed, it began to look more and more as if the interpretation would hold water. Now this is not to say that there were no problems. For example, the experiment had not been conducted terribly well in terms of using the latest up-to-date equipment and measuring devices. (In fact, the Bulgarian in a footnote thanks his brother—an Australian sheepherder—for the loan which allowed him to buy the war-surplus electric motor with which to carry out the experiment. The motor, while it worked, did not work all that well compared to what a well-equipped contemporary laboratory might accomplish with its lasers, jewel bearings, and computers.) But even given his crude equipment, the Bulgarian had made a point, and made it well enough for a genuine American establishment scientist to take notice.

This story seems to provide some counterevidence to the claims of the politicians. In fact, it raises at least one severe difficulty for their view. One important aspect of the politician theory is that the "politics as usual" condition *must prevent* any nonpolitically well-set individual from even the first step toward acceptance; this first step being getting his foot into the door, the logical precondition for theory acceptance. It is a logical precondition because, if you cannot even get your foot in the door, then you will never be able to get your theory evaluated in terms of its acceptability. Thus the politician view must rule out even initial consideration if it is to accomplish what it is supposed to, namely, the political explanation of how theories come to be accepted/established. But this present example casts doubt on this aspect of the politician view. Consider the case more thoroughly. Clearly, even though the Bulgarian had two political strikes against him, namely, his isolation from the big leagues of science, and his totally unfashionable hypothesis, he still somehow managed—contrary to politician beliefs—to get his foot in the door at least to the point of having his experiment carefully scrutinized by an established authority in the field. I must also emphasize that, although this is only an individual case, there are enough cases which illustrate similar situations to give pause to the strictly political view of theory acceptance.

There is a sequel to this story which helps to put the political view into the proper perspective. My friend has described for me what he will now have to do in order to get a relatively fair hearing for the Bulgarian's view. He certainly cannot, in any circumstance, come right out and announce, "I have here some results from a Bulgarian scientist which indicate that Einstein's views about absolute space are wrong." Were he to do that, the political machinery of the scientific institution would make extremely short work of grinding both the Bulgarian and my friend himself into very fine grist. No, what he has to do is to present the results as an oddity, as an apparently anomalous situation which is in dire need of straightening out by the experts. Thus, we might expect him to say something like, "Here are some assuredly wrong, but curious, results; we might run the experiment under better technical conditions—if anyone can find the time and interest to do so." With this set of tactics, I cannot predict anything but success for my friend. He will hit the scientific community in the soft underbelly, the only spot completely unprotected by political defense—their curiosity. This approach will certainly bring results.

The Ossiander Maneuver The tactics involved here are legendary in science, most likely because of one of their most notorious historical precedents. I call these tactics the *Ossiander maneuver* in reference to that notorious precedent. Ossiander was Copernicus's good friend and editor. Copernicus had spent years working up his theory that the earth went around the sun, instead of vice versa. The earth-centered or geocentric cosmology had been around for 1800 years before Copernicus. It had gotten itself all tangled up in religious beliefs, literature, politics, and every other facet of human intellectual existence. To put it bluntly, the geocentric view was a paradigm of scientific paradigms. And here was poor Friar Copernicus, about to attack this fortress of belief with nothing else than some geometrical considerations and a couple of well-reasoned arguments. Moreover, to make matters worse, it became evident as one plowed through Copernicus's book (its reading difficulty was a legend even in its own time) that Copernicus *himself* had come to firmly believe that the earth circled the sun, rather than the opposite. Ossiander, realizing full well what might happen—indeed, what was *guaranteed* to happen—if Copernicus published the manuscript as it was, tried to head off trouble before it came to pass. In short, what he did was to write a preface in which he asserted a complete fictionalist position: According to Ossiander, Copernicus did not *really* believe that the earth circled the sun, not at all; rather, what was being proposed was merely a convenient fiction, a nice, logically elegant fairy tale, one which made the computations of the navigator and calendar maker just a little bit easier. So said

Ossiander, and thus was history made. Ossiander's move worked. Copernicus's book was read, and it was understood; only then was it attacked. But by the time of the attack the important seed had already been planted, and indeed it had germinated. Once the seed of the heliocentric theory had germinated, it was only a matter of time before the likes of Kepler and Galileo came to accept it and what it taught about the new shape of our concepts of the universal system.

It seems to me clear what this comes down to. It is quite true that science, just like any other human institution, has its manifestly political aspects. If you come up at cross-purposes to these political currents, your hypothesis stands little likelihood of ever being accepted. In fact, your hypothesis stands little chance of ever being initially considered. However, to stop the account at this point, in full agreement with the politician's view, is to do a complete injustice to the full reality of the scientific institution. In the first place, it is obvious that the political aspects of science do not eliminate our hopes for its rationality; that is, merely because science requires political acumen it does not follow that science is antirational. Popper's hope for an open society will not necessarily founder upon this rock. Rather, the rock must merely be avoided. The feasibility of the Ossiander maneuver, which is open to any scientist who wishes to propose an unorthodox view, and/or to do his proposing from the Bulgarian, or the Transylvanian, or even the Outer Mongolian research station, shows that politics is merely *an* aspect, but not the whole, of science. If the Ossiander maneuver can be successfully brought off, then the seed, however tiny, will have been planted. From that point onward, pure scientific curiosity will motivate the eventual decision on the acceptability of the hypothesis.

The Lysenko Case Politics has other sorts of influences in theory acceptance. An example is provided by another curious case where politics played an active role in science. Moreover, in addition to being a political example, it is a case which shows the futility of a strict idealist position as a philosophy of science. The situation involves a man named Lysenko. Lysenko was a Russian geneticist who, for various reasons, came to be the commissar of all Soviet research in genetics and applied plant/agronomic science. His decision came to be the final word on which hypotheses were acceptable, which experiments were to be funded, and so on. There is not much problem with this situation—or not yet; Western scientific institutions show more and more signs of becoming centralized in just the same way. Centralization itself is not the problem. The problem comes in when we consider what were Lysenko's own personal biological beliefs. Lysenko was a quasi-Larmarckian insofar as he believed that the life experience of a living entity could have influence upon its progeny. In this way, he believed, acquired characteristics could be taken into the genetic stream of a domesticated plant or animal race. But even more important than his specific beliefs was the fact that Lysenko was a complete and total dogmatist: He *really* believed in his theories, and he had the power to compel belief, or at least pretended belief, among his colleagues and subordinates. For some twenty-five years Lysenko diverted biological research into his own chosen channels, channels which were congruent with his paradigm theory of the genetic world. Needless to say, Soviet science fell far behind during Lysenko's tenure. Genetic breakthroughs which had been painstakingly brought about by Western scientists were ignored by the Russian establishment. Lysenko's political control was constantly used to maintain this condition. In terms of our philosophical analysis, what came about through this complete politicization of science was an effort to bend the natural world to fit the theory which existed in the concepts of the political boss of the science. Such a method is complete idealism, that is, the effort to make the world conform solely to our ideas about it. It did not work. The world resisted in the end, and Lysenko's ideas could not be forced upon nature beyond a certain level. Ultimately, to make a shorter story of it, Lysenko fell from power. Since then, there has been an accelerated effort to catch up, and Soviet genetics is no longer isolated from the tenor of modern times.

Again, however, just as in the case of the Bulgarian, this case does not show that the politician view of theory acceptance is the correct one. Rather, it again indicates the resilience of the scientific conceptual structure—the dynamics of scientific paradigm and tradition—in the face of all-too-human foibles. Politics is certainly a *part* of science, but it is neither the most important part nor a sufficient part. The worlds of the scientist, both natural and social, ultimately resist the complete politicization of the scientific society. Whatever the correct view of how the theory-acceptance process works, we must say that the political aspects of it are only elements in the process, and not the whole process itself. The politician view is accurate as far as it goes, but we ought not to go any further than that in espousing it.

The Sociologists

Even though the sociologists do not spend much effort trying to study theory acceptance in itself, some of the topics which concern them do have some interest for the topic. Sociologists have been most concerned with trying to discern the institutional structure of science, that is, to examine science as a society, complete with all normal societal features—leaders, communities, channels of communications, systems of award and sanction, and so on. Science, for various reasons, a main one being its formalized procedures, is a very nice object for sociologists to study. The Nobel prizes, to take just one example, are a clean case study in the awarding of status in a community.[17] In another nice example, citations—research paper footnotes—are an apparently precise means for cataloging intellectual influence. And other examples could be cited.

In a fashion similar to the politicians, the sociologists do not seem to believe that their task is to specify when a discovery ought to be accepted; rather, they have busied themselves in trying to find criteria which will allow them to decide that *in fact* a discovery has been accepted. Again, they are interested solely in describing how science itself works, without attempting in the least to prescribe how it ought to work, or when it is working well or badly, etc.

The proposed sociological criteria for deciding when a theory has been accepted can be relatively straightforward. One notion, for instance, is quite similar to that of the politicians. The sociologists involved believe that science is a society which is organized in such a way that there are "leaders"—the well-renowned scientists, the directors of major experimental laboratories, the editors of major journals, and so on. Theory acceptance according to these thinkers is simple to define on this model: A hypothesis is accepted if and only if the leaders accept it. Since the leaders are indeed *leaders*, then, if they accept an idea, ipso facto, it is accepted. Unfortunately, even given its intuitive reasonableness and plausibility, the problem with this view is clear: Science probably does not have leaders in the required sense. It is quite obvious that every scientific community has mavericks, loners, and all other sorts of persons who would count, in this view, as societal misfits. These individuals clearly are not the types to play "follow the leader" in the fashion required by this particular sociological view. But even more significantly from the point of historical accuracy, most of the major discoveries in modern science were initiated by mavericks—or, at least, individuals who were perceived as mavericks when they were viewed against the more stable backdrop of the paradigm/tradition they were confronting. This after all is the whole point of attempting to understand paradigm change as "revolution." Men like Lavoisier, Pasteur, and Einstein came out *against* the leaders, those eminent scientists like Priestley, Liebig, and Newton, already in their initial pronouncements. This sort of behavior simply does not fit into the neat sociological model provided by the notions "leader" and "follower." Thus this particular sociological model simply fails to describe the actuality of scientific history.

A second view is even more straightforward. Unlike the "leadership" view, it attempts to provide absolutely no explanation at all of theory

[17]Harriet Zuckerman, "The Sociology of the Nobel Prizes," *Scientific American*, November 1967, p. 25.

acceptance. It simply provides an operational definition which allows the social scientist to identify a theory as being accepted or not. According to this definition theory *x* is accepted if and only if it is a significant element in new textbooks. When some new hypothesis, for example, the continental drift hypothesis, suddenly becomes all the rage in the textbooks, then it is indeed safe to conclude that the hypothesis has been accepted. Continental drift theory, if one would care to examine the case, hardly rated a mention in the geology texts of the 1950s. But by the mid- and late 1960s no text could be published which did not have it as a main focus of explanation. Because of its fit with historical data such as these, the "textbook" criterion clearly provides an accurate basis for deciding when a hypothesis has been accepted.

Unfortunately, its accuracy is its only saving grace. Otherwise, the textbook criterion does not provide us with any information at all about theory acceptance. It leaves all the interesting questions unanswered (indeed, it leaves them unasked). Consider, for example, the following puzzles: How do textbooks come to be written? By whom? Who decides which textbooks to use? On what basis is it decided whether or not a textbook is a good one? And by whom? What role does the publisher play in all this? How about the economic issues, such as, what relation does the price of a textbook bear to its popularity? And so on ad nauseum. It is quite clear to me that the textbook criterion simply does not have enough meat to figure in any serious attempt to analyze theory acceptance. However, in conjunction with other views, perhaps the textbook criterion can be used at least as a device to start investigation into the theory in question. That is, since it does its job accurately, it provides a solid jumping-off place for further historical, philosophical, and sociological analyses.

SUGGESTIONS FOR FURTHER READING

Although much has been done since its publication, Mary Hesse's *Models and Analogies in Science* (Notre Dame: University of Notre Dame Press, 1966) remains an extremely useful introduction to the subject. Aesthetic principles are a very murky subject, but G. Schlesinger has managed to say quite a few clear things on the topic. See his *Method in the Physical Sciences* (London: Routledge and Kegan Paul, 1963); chaps. 1 and 3 especially have nice discussions of the methodological use of aesthetic principles. L. Susan Stebbing's *Philosophy and the Physicists* (New York: Dover Publications, Inc., 1958), pts. I and II, have a very nice bit of plain thought on the fictionalist versus realist (extreme and otherwise) issue. On the problem of theories of truth, I especially like chaps. 1 and 2 of Nicholas Rescher's *The Coherence Theory of Truth* (Oxford: Oxford

University Press, 1973). As always, Rescher is somewhat tough to read—but again as always, it is worth the struggle. Political currents in science have been especially vicious and frightening at times. One of these times centers in the Velikovsky affair. See "The Politics of Science and Dr. Velikovsky," *The American Behavioral Scientist,* September 1963. Since my treatment of the social science analysis of theory acceptance is so brief, the student who is interested in some further reading of these analyses would do well to consult, for example, D. S. Greenberg's *The Politics of Pure Science* (New York: New American Library, 1967) or J. R. Ravetz, *Scientific Knowledge and its Social Problems,* (Oxford: Oxford University Press, 1971). Ravetz in particular argues that dogma has a positive function in scientific change (chap. 9). W. vO. Quine also points out quite correctly that any scientific statement whatsoever may be tenaciously clung to in the face of opposing evidence. This point relates directly to both the problem of dogma and the logic of the Duhem/falsifier controversy. See W. vO. Quine, *From a Logical Point of View* (New York: Harper Torchbooks, 1961).

Chapter 9

Acceptance of Lavoisier's Hypothesis

INTRODUCTION

In 1772 Lavoisier proposed that the weight gained by phosphorus and sulfur during combustion resulted from the "prodigious quantity of air that is fixed during combustion" and that enters into combination with the original substance.[1] The observable facts of the situation, as we have seen, were expressed in the mere correlation "*x, y, z* burns" and "*x, y, z* gains weight." Aligned against Lavoisier's hypothesis was an exactly opposite theory, the phlogiston theory, which postulated that the correlation "*x* burns" and "*x* gains weight" was the result of the mechanism "*x* loses phlogiston during combustion." Thus the two opposed hypotheses were (1) Lavoisier's view that "*x* combines with something from the air during combustion" and (2) the prevailing view that "*x* loses phlogiston during combustion." In addition, we must remember that the phlogiston hypothesis was not directed solely to these facts about combustion and calcination; phlogiston figured as well in explanations about smelting,

[1]James B. Conant (ed.), *The Overthrow of the Phlogiston Theory*, (Cambridge: Harvard University Press, 1966), Case 2.

human physiology, and in general, any phenomenon involving heat and energy. In terms developed in Chapter 8, we are forced to admit that Lavoisier was up against a well-established concept structure, one, moreover, which provided quite a bit of internal coherence for the various observable phenomena to which it was directed. Lavoisier's job was cut out for him, and it was no easy task for him to get his own hypothesis established in the face of such well-entrenched opposition. How he did so is a marvel of scientific achievement. Let us examine his moves.

My analysis of Lavoisier's tactics will focus upon two separate but related aspects: First, Lavoisier must settle the *existence* question; second, he must settle the *conceptualist* question. The first of these aspects is a necessary ingredient in Lavoisier's work for a very simple reason. Lavoisier's original hypothesis was about a certain "something," an *x* which combined with combusting substances. An inescapable ramification of his postulating such an *x* is the finding of it, the "bringing-back-alive" of it. There is no use in merely postulating the existence of something but then leaving the matter at that; it is simply unavoidable that a search must be carried out to find the new "beast." Postulation of a new object involves its existence; Lavoisier's actions subsequent to 1772 bear out the accuracy of this general rule.

In addition to bringing his *x* back alive, Lavoisier had another and even more difficult task. Since his hypothesis was up against a well-entrenched, highly coherent conceptual system, eventual acceptance of his proposed new concept system was totally dependent upon his making it as coherent as the system it would replace. As we shall see, Lavoisier was able to do just this and, moreover, to add to the requisite internal coherence between observational phenomena a further, external coherence between his hypothesis and the physics of his time. I will argue that this last feature of Lavoisier's work is very significant, especially since it involves the exploitation of a consistent *quantificational* scheme for the first time in chemistry. However, before discussion of the conceptualistic issues, we must settle first things first, namely, the problems with the existence of Lavoisier's mysterious "something."

THE METAPHYSICAL ISSUE

Lavoisier's hypothesis about the mysterious "something" had been contained in a sealed note deposited secretly with the secretary of the French Academy. Although this procedure might seem a bit strange, it was a typical thing to do in the early days of science. When some investigator thought that he had created something new and potentially valuable, but had not worked it out thoroughly enough to publish it, he usually made rough notes about it and left the carefully sealed notes with the secretary of his scientific society. This move not only protected his priority in the discovery, it also protected him from ridicule if the idea turned out to be worthless upon further research. In Lavoisier's own case, his caution paid off. He experimented for the next two years trying to discover the nature of the mysterious "something" which was combining with the sulfur and phosphorus during their combustion. He went so far as to report his experiments, but it was clear that they just plain did not lead anywhere valuable. In some respects, indeed, his results with sulfur were wrong. But he did achieve one thing, namely, some success in his work with the balance. Also, his techniques for collecting gases and weighing them improved quite a bit. His reports on all this work set it up so that his colleagues were beginning to become more familiar with his efforts, especially his penchant for the quantitative. Yet it was clear to all that his work was not achieving any astonishing success during this period.

The reason for this lack of success is fairly straightforward. Lavoisier was up against a common chemical problem: Chemical analysis depends intimately upon the purity and identity of the substances under investigation. On the other hand, the purity and identity of the substances depends necessarily upon the correctness of the chemical analysis. I think that you can see the "chicken-or-egg" circularity of the problem. When chemists attempt to discover the natural properties of substances, it is important that they know whether or not the sample under investigating contains significant amounts of other substances, amounts which could throw off the interpretations of the results. If this is not known, the results are unreliable. Let me give you an example of how this might work.

Archimedes

Almost all substances which occur in nature are in some sort of combination with other substances. Even gold, which is sometimes found almost pure, is usually contaminated by trace amounts of silver and lead. If chemical analyses are carried out on this naturally occurring gold, it is likely that the results will be complicated by the chemical side reactions with the traces of lead and silver. But discovering whether or not there are adulterating trace amounts presents a serious problem, a problem of discovery almost as difficult as those involving theory creation itself. The story of Archimedes and the golden crown well illustrates the difficulties of the problem. Archimedes' king had given a goldsmith a certain amount of gold in order that a crown might be manufactured. When the finished crown was delivered, it happened that the king was suspicious that the smith might have kept some of the gold for himself, and mixed in small amounts of lead and silver, which were quite a bit cheaper in price. In this way the smith could make just a bit more profit off the deal. The king summoned Archimedes and bade him to invent an analytical test which could determine whether or not the gold had been adulterated. Archimedes spent a long time thinking about the problem (one does what the king requests . . . or else!), but he just could not come up with a way—especially a method which would not destroy the crown. Although there are various chemical tests we now know about which would have worked, Archimedes in 400 B.C. certainly could not have been expected to know that a gold sample dissolves in a different fashion in aqua regia acid than does a sample of lead-silver-gold alloy. Archimedes had none of these subtle chemical tests at hand. Eventually, however, the famous "Eureka!" experience happened. Archimedes jumped in the bath, noted (somehow) the relation between the volume of the water his body occupied and its mass, and from this relation created his law of buoyancy. As it turned out, he had discovered the concept of *specific gravity* which broadly predicts that every substance has its own peculiar mass per unit volume. According to this analysis, a gold alloy would have a different mass/volume ratio than would a lead-silver-gold alloy. Although I have completely glossed over the difficulties, the incredible puzzles Archimedes was facing, I think that you can still see the problem here: In order to find out what he had in the crown, Archimedes had first to figure out a way to figure out what he had in the crown. That is, he had to have at least a rough idea beforehand what were the potential substances in the crown, their range of qualities and interactions, and possible ways to distinguish among them. Archimedes had to rely upon certain physical facts which he already knew in order to discover what were the facts in his new case. He already knew that most substances have a roughly specific mass per unit volume, but the main problem was how to measure the volume of the crown. Using water displacement as a volume-measuring technique is a stroke of genius, indeed, it is Archimedes' very own stroke of genius. It must be pointed out, however, that Archimedes' technique is purely physical; that is, it relies solely upon physical dimensions such as mass and volume.[2]

The chemical sort of case faced by Lavoisier and his fellow early chemists is even more tricky than the physical kind of case faced by Archimedes. There are at least three kinds of special difficulties. First, most chemical substances under investigation by Lavoisier and his colleagues were more or less contaminated by other substances. Second, and this is especially important from a conceptual point of view, the notions of "mixture" and "chemical compound" had not yet been completely sorted out. That is, Lavoisier and his contemporaries were not

[2]I suppose, by the way, that you would like to know how it all turned out for the goldsmith? Well, so the story goes, the crown *had* been fiddled with. Archimedes figured it out and the goldsmith paid dearly for his attempt to cheat the king.

yet explicitly aware of the precise difference between two substances being merely physically mingled, e.g., a mixture such as salt and sugar (or gold and lead, for that matter), and substances which are chemically bound together, such as the chlorine and sodium in table salt. Finally, to make matters worse, a significant amount of their work was concentrated upon gases–which are for the most part invisible, odorless, tasteless, tricky to collect and handle, and in general extremely difficult to deal with.

All these problems came together in a particularly vicious way in Lavoisier's work. Take the purity problem alone. Lavoisier ended up doing work with calxes, substances which, as we have seen, are intimately related to the ores used in smelting. But the calxes he had available were invariably mixtures. A classic case involves the various calxes of lead. Lead substances had been used in tinting and painting for a long, long time. In most cases, the specific color of the lead calx is a reliable clue to its identity—or, rather, to the identity of the constituent substances which make it up. Thus, a red lead calx would be assumed to be composed of different substances than would an orange or yellow lead calx. But as we now know, the different colors are not necessarily related to compounds of different underlying substances. Rather, the different colors might very well *not* be related to a difference between the kinds of substances, but merely to differences in the chemical structural relations between two substances that were exactly the same in other respects.

Then take the gas problem. Smelting had become known to involve the giving off of a gas during the reaction. After an awful lot of work, this gas had been identified as being the same as that which was given off during combustion, brewing, and breathing. It was known as "fixed air" and could be positively identified through several specific tests, viz., it dissolved in water upon agitation, it would not support combustion, it turned limewater cloudy, and so on. These essential facts were tied together by the hypothesis that "fixed air" was regular air which had become saturated by phlogiston, which itself was given off during combustion. But fixed air was not the only gas known. The British chemist Joseph Priestley (you will soon find out much more about him) had discovered at least three other sorts of "airs," all of them involving combinations of regular air—or one of its parts—and something from nitric acid. As we know now, Priestley's three new airs were compounds of nitrogen and oxygen. But oxygen and nitrogen are devilishly complicated in their interactions. They can produce a whole series of different chemical structures, each with its own peculiar idiosyncracies of behavior. One of these airs of Priestley was reddish in color, and dissolved in water; yet another was colorless and unsoluble, and would not support combustion; while the third was colorless, was slightly soluble in water, and most importantly, supported combustion.

The point of all this is simply to try to indicate to you the incredible complexity of the situation faced by chemists. Everywhere they turned they found messy, complicated, impure, real-life situations which they had to try to sort out. The major problem of course is that sorting out the situations required having the correct concepts, and as I have argued many times before, the correct concepts must both precede and follow the work of identification and definition of the substances involved. But Lavoisier was not without a bit of good fortune in this matter. He had at least two things going for him. In the first place, he was able to procure his experimental substances from one of the best pharmacists in all of Europe. This man, M. Cadet, had an enviable reputation for the purity of his drugs and chemical substances. Hence, insofar as purity could be assured, Lavoisier was assured about his chemicals. But the second fortunate circumstance is even more important.

Of all the substances which can be involved in combustion, calcination and smelting, there is one—and only one—which can be dealt with simply, and at low temperatures. By "simply" here I mean that it can be acquired in a relatively pure state in nature, or at least refined into purity unproblematically, and moreover, it can first be decomposed and then subsequently *rebuilt* with a fair amount of ease. Finally, its most important feature is that it can be smelted without the addition of charcoal. This is quite significant because it allows the researcher to work with a smelting reaction that is simpler by at least one component. The substance I refer to here is *calx of mercury*; it is a red powder, obviously earthy in its qualities. Its unique chemical properties allowed rapid progress to be made. Although serious work on it did not begin until early 1774, already by late 1775 its features had been unraveled to the point that its behavior in calcination and smelting was nearly understood. Then, jumping off from this point, using his understanding of calx of mercury as a paradigm, by 1777 Lavoisier had produced a complete, totally new conceptual system for combustion, calcination, and smelting which was a serious candidate to replace the phlogiston theory. It is difficult to overrate the significance of the peculiar of behavior calx of mercury. The history of chemistry would clearly have been different were it not for this unique substance.

We shall now leave the methodological points and look more closely at the details of the historical events themselves. Before examining the details, I will sketch in the broad outlines of the story.

Priestley, the English Phlogistonian

Joseph Priestley, the British scientist, believed in phlogiston theory. He also worked on calx of mercury; indeed, in some ways his work on this substance was superior to Lavoisier's. Priestley's work disclosed Lavoisier's "something," the mysterious *x* from the air. But Priestley did not recognize this *x* for what it was to be (it would have been impossible for him to do so, since his idea about the identity of the *x* was wrapped up in phlogiston theory). However, Priestley apparently shared his news about the discovery of the misidentified *x* with Lavoisier. Lavoisier for his part had no difficulty at all understanding that his mysterious sought-for *x* had been found; thus he immediately tried to run the relevant experiments. Unfortunately, at this point he made a mistake in his interpretation, but he published the results without realizing his error. Priestley meanwhile had continued his work, and was now in a position to correct Lavoisier's mistake, which he soon did. Lavoisier, acknowledging the error but not Priestley's help, finally got the identity of his *x* right, changed his experimental report, republished it, and thus established the internal and externally coherent theory of what he later called "oxygen." Priestley never did accept Lavoisier's theory, even though, late in his life, he admitted that Lavoisier was most likely correct in his new system of concepts. Thus evolved the events of an exciting set piece in scientific communal enterprise. Let us look at the details.

Priestley was in his early forties at the time of his confrontation with Lavoisier. He was an acknowledged scientist, well known for his eclectic but careful dabbling into just about anything and everything. Like most British scientists of the period, he was not a professional in our modern sense. Rather, again somewhat typically, he was a clergyman with a regular pastorate.[3] But his clerical duties left time for him to carry out his researches, and he was thus able to follow his curiosity wherever it led him. Priestley, in addition to his meticulous chemical labors, also did quite a bit of very clean and careful work in electricity. His publications in this field were very significant, especially in their influence upon people like Faraday and Franklin. But it is his chemical work which interests us now.

Priestley had come early to a hypothesis that the air we breathe is not a simple substance, but that it is a compound of at least two and possibly three other substances. Although this does not sound like an especially daring proposal, we must remember that the traditional view during Priestley's time was that "air" was a simple element, this belief being an item in the concept system which had evolved out of the four-element

[3]The connection between the natural sciences—geology, chemistry, and biology in particular—and the religious life in Britain during this period is a fascinating one. Although I certainly cannot go into any detail about it here, one can easily discern the economic and sociological situation which underlay the connection. Britain is filled with small villages and parishes, each of which has associated with it an established church and its "living." Many scientifically inclined college students went into the clergy, since were they to secure a living in a suitable parish, they could thus support their scientific researches in a somewhat leisurely fashion. Someone someday must tell this story about how the English church in effect supported scientific research in the eighteenth and nineteenth centuries.

(air, earth, fire, and water) paradigm of the ancients and medievals. Priestley's work during this period can be understood then to be related to his trying to establish his hypothesis about the nonsimple nature of the air. In aid of this research he had been forced to develop methods of collecting gases, techniques for storing them inertly, and ultimately, tests which could qualitatively identify them. Even though his "complex-air" hypothesis was a bit revolutionary, his methodology was not at all out of line with the phlogiston/four-element tradition. His work *was* qualitative; that is, he was searching for methods of identifying the qualities, the peculiar properties and features, of the substances involved. So right from the start his work was classical and traditional in a manner directly opposed to the quantitative techniques which were even then being developed by Lavoisier. The fascinating thing to my mind is that each of the two men each needed the other and the other's divergent views. Lavoisier, as it turns out, made better use of Priestley than vice versa.

Calx of Mercury

Enter now the calx of mercury. The red powder had been decomposed first by the French chemist, Bayen. He heated the powder alone, and collected the "air" which was given off during the heating period. But when he tested the sample, he somehow made the incredible error of identifying the evolved gas as "fixed air." Some historians have been quite skeptical of Bayen's reports, and given the straightforward and reliable tests for fixed air—tests which everyone knew and could use with facility—it *is* rather difficult to understand how Bayen's error could have been so drastic. But in any case he publicly reported that the calx could be dissociated into pure mercury and fixed air. The import of this experiment, even flawed as was its interpretation, was obvious. Since pure mercury metal was produced by the reaction, it was evident that the reaction was a *smelting* reaction. This made it potentially explicable in terms of phlogiston theory. Indeed, this new set of observable data is easily fitted—although not without a certain amount of puzzlement—into the phlogiston system of concepts. Consider the paradigm phlogiston explanation of smelting:

$$\text{Calx (ore)} + \text{charcoal (phlogiston)} \Rightarrow \text{metal (phlogisticated calx)}$$

According to this general equation, during smelting an ore has added to it some phlogiston, which comes mainly from the charcoal. It is heated in a fire—which is also rich in phlogiston—and the result, metal, is produced. The general equation can be specifically applied to the case of calx of mercury:

Calx of mercury + phlogiston $\Rightarrow$ metallic mercury

The puzzling feature was that no addition of charcoal was necessary. The puzzle, however, was only a minor one, at least to convinced phlogistonians. After all, there was an abundant source of phlogiston in the fire itself. It is easy to explain away the puzzle by conceiving that mercury calx is just a slightly variant calx, a peculiarly idiosyncratic ore, which does not need quite the overabundance of phlogiston supplied by the charcoal in order to turn into its metal. Moreover, Bayen's report that the residual gas was fixed air was quite coherent with phlogiston theory, since all other smelting reactions also evolved fixed air during the process. Thus, with the exception of the minor idiosyncracy, the calx-of-mercury reaction ties in nicely with all the other smelting reactions covered by phlogiston theory.

Priestley soon disrupted this appearance of order. Within six months of Bayen's work, he himself had duplicated the noncharcoaled mercury smelting. It was immediately obvious to him that the evolved gas was not fixed air. For one thing, the gas supported combustion. Second, it did not produce a cloudiness when mixed with limewater. Failure of these critical tests provided an absolute refutation to the notion that the evolved gas was to be identified as fixed air. Priestley thus corrected Bayen's results. Note that this event is a very nice illustration of the communal, public nature of science. Bayen made his report public and, in so doing, put himself forward as a target to be shot at by the other members of his scientific community. Priestley then duplicated Bayen's experiment, reinterpreted it, and took a good shot at Bayen, a shot which scored. Bayen's notion thus failed the trial by fire.

Unfortunately, even though he showed that the gas was not fixed air as Bayen had thought, Priestley was not correct in his own identification of the gas. He thought that the gas was one of his three nitrogen airs—the odorless, colorless nitrogen air which supported combustion. (Misidentifying the mercury calx gas seemed to be the rule of the day, since Lavoisier, within a year, was *also* to misidentify it.) It is easy to suppose that Priestley's attachment to his own "airs" had let him down in this case, which is one obvious deficiency in the methodology of "hypothesize first, then experiment" which I have been arguing for. In any case, Priestley got the gas wrong. This was in August 1774.

The next event has proved to be a nice one for historians of science. In November of that same year, Priestley went to Paris. He spent a lot of time in the company of the French chemists, and Lavoisier in particular. According to Priestley's own report of the trip, he mentioned his work with calx of mercury to Lavoisier, and went into some detail about the gas which resulted from the smelting. Lavoisier himself never really reported these conversations in detail; moreover, he certainly never did give any credit to Priestley for putting him on the trail of the red calx. Thus, we have only Priestley's version of what had been discussed. But, for whatever reason, Lavoisier started doing his own crucial experiments on calx of mercury within the month. Historians have since spent a lot of delighted effort trying to trace down the exact details and events of these contacts between Lavoisier and Priestley, but no conclusive data have been found. However, it seems quite safe to infer that Priestley's information must have had *some* beneficial effect upon Lavoisier's research program. Lavoisier's initiation of calx-of-mercury experiments so soon after Priestley's visit is just too pointed to be a mere coincidence.

Lavoisier's Experiments

Lavoisier's experiments were reported in a paper he read to the French Academy on Easter 1775. Even its very title shows its connection to his original hypothesis of 1772: "On the Nature of the Principle [substance] which Combines with Metals During Calcination and Increases Their Weight." Lavoisier, in a series of experiments with the red calx, had been extremely careful to account for all the weights of the substances involved. Moreover, he had run the smelting both *with* charcoal and *without* charcoal, and in each case accounted for the weights of all substances both before and after the interaction. His doing this shows some really nice logic, which we need to unpack just a bit in order to appreciate its significance.

Lavoisier first did the run using charcoal. The results were exactly as conceived in phlogiston theory: The calx and charcoal were entirely consumed, leaving only metallic mercury and fixed air in the final product. The beginning weights and the ending weights were exactly identical. But Lavoisier proposed that the fixed air was not a simple element; rather, he said, let us conceive that the fixed air is a compound of the charcoal plus one of the substances which compose the calx. Already this hypothesis is in complete and absolute contradiction to the phlogiston theory. Lavoisier was in fact proposing that the calx was not a simple element, but rather was compound in nature—an interpretation just the exact opposite of the phlogiston conception. Moreover, if the calx was a compound, then the mercury metal was a simple element. Finally, if this was the case, then the reaction was *not* a combination reaction between the calx and phlogiston, but actually was a dissociation reaction in which some underlying substance was stripped away from the calx. The basis for this proposal is to be found in the second stage of his experiment, the run which was made without any addition of charcoal. In this version of the reaction, Lavoisier simply took the red powder, heated it long and hard, and then weighed the resulting metal and gas. The weights added up neatly. This means that nothing, *no mass, was added to the reactants during the experiment.* Thus, although phlogiston theory predicts that phlogiston is added to the calx in order to smelt it to the metal, Lavoisier's results indicate that the metal weighs less than the calx (which implies that phlogiston has a negative weight), and most importantly, that the weight lost by the calx is just exactly identical to the weight of the residual gas. Moreover, and this is the final clincher, the gas which is evolved in the smelting done without charcoal is most definitely *not* fixed air. Lavoisier showed this last point with an excruciating attention to detail: He showed that the gas would support combustion, it would support animal respiration, it did not turn limewater cloudy, it was insoluble in water, etc. Thus, whatever the gas was, it was not fixed air.

From this point onward, however, Lavoisier himself fell into error, the same error earlier made by both Bayen and Priestley: He misidentified the gas. Lavoisier's misidentification is an interesting one. He believed that what he had generated by the reaction was none other than the common air itself, whole and entire. In fact, he noted, the "air was not only common air but that it was more respirable, more combustible, and consequently that it was more pure than even the air in which we live."[4] On this interpretation, then, calx of mercury was a compound composed of ordinary air plus mercury. When heat was applied the compound came unstuck, leaving the mercury in its liquid, metallic state with the air floating above it in gaseous form.

Underlying Lavoisier's mistake is a technique earlier developed by Priestley. Priestley had worked up a method called the "goodness test" for checking the purity of air. This test consisted of blowing a quantity of one of Priestley's nitrogen "airs"—this time the odorless, colorless, water-insoluble one—into the sample of air he wished to check. This original quantity was measured, as also was the quantity of the sample. If the air in the sample was still "good" (respirable, supportive of combustion, etc.), then there would be an immediate reaction: A red gas would be formed, and this gas would go into immediate solution with the water in the base of the collection jar. Priestley then would measure the volume of the gas which remained after the red fumes had dissolved, and this would provide a measure of the "goodness" of the air involved. Lavoisier used this test to analyze the "air" which came off the heated calx, and its volume was quite close (certainly within the limits of experimental error) to what it would have been according to the standard test with common air. The results of the "goodness" test, conjoined to the results of the other tests, apparently offered conclusive proof that the gas was indeed common air. And this Lavoisier concluded.

It was all a mistake, of course. Through an unlucky accident, the

[4]Ibid., p. 27.

volumes originally chosen by Priestley as standard for the goodness test just coincidentally happened to match the volumes which Lavoisier used in his experimental setup. Thus, no matter how carefully Lavoisier had run the test, the accidental match-up of the volumes would mislead him into thinking that the gas was common air, and not some part of it. In the 1775 Easter memoir, then, Lavoisier thought that he had identified the mysterious "something from the air which combines with metals during calcination and increases their weight": The "something" was none other than the "air itself entire without alteration."

Priestley's Experiments

But Priestley had not been sitting around idle while Lavoisier was duplicating his experiments. He himself was duplicating them as well. During March a chance event led Priestley to make a more correct identification of the substance given off as a gas from heated red calx. Priestley had secured a giant magnifying glass—a "burning lens" as he called it—and he used it upon the sample of calx he had secured from M. Cadet in Paris (it probably came from the exact same lot as did Lavoisier's). He smelted the calx under the lens, and produced the gas in question. He then performed the goodness test upon it, and got the same results as Lavoisier. He went to bed, slept, awoke, wandered back into the lab, and then, as he describes it, for no apparent reason he thrust a burning candle into the collection vessel where he had the day before run the goodness test. To his complete and utter amazement, the candle burned, indeed it flared, for another considerable time. At this point, the true nature of the mysterious *x* from the air was open for disclosure.

The candle should not have burned. When the goodness test was run on normal air, it completely exhausted the respirable principle of the sample. Candles put into the remainder went out, burning splints immediately extinguished, and living animals quickly went unconscious. But Priestley's morning-after candle did not die; rather, it lived happily for a long period of time. What this indicated to Priestley was straightforward: The air given off from the calx of mercury was better, of a higher degree of "goodness," than garden-variety air. Priestley decided immediately to see just how much better it was. He reran the experiment, collected another sample of the gas, and pumped in his nitrogen air until the red fumes quit forming. As it turned out, the calx-of-mercury air was able to absorb four to five times as much nitrogen air as did ordinary air. Thus the calx's air was four to five times purer than ordinary air. What could this stuff be? Priestley answered the question quickly.

In terms of the phlogiston theory, phlogiston is given off during combustion and respiration. Combustion and respiration cease when the air reaches its saturation level of phlogiston. This is usually after about 20 percent of the available air has been used up. Accordingly, we might say then that ordinary air is about 80 percent saturated with phlogiston in its natural state. But what about Priestley's new calx-of-mercury air? Since it was about four times more pure than ordinary air, this could be interpreted as meaning that it was air which was 100 percent unsaturated by phlogiston. So Priestley thus named it in a very coherent, phlogistonian manner; the air given off during smelting of mercury calx was to be called "completely dephlogisticated air." Far from being common air, as Lavoisier had thought, the calx-of-mercury air is nothing but the best, most eminently respirable air; moreover, it is an independent substance in its own right, one which provides one of the stuffs which mixes in with others to produce common air.

Priestley did this work in March. Lavoisier's memoir identifying the gas as "common air" was delivered in April, and printed shortly thereafter. By November 1775, Priestley had corrected Lavoisier's mistake—in print. At this point it was all over. Priestley had isolated the stuff, shown how to test for it effectively and correctly, and corrected Lavoisier's misidentification of it. But Priestley then went conservative and stayed entirely within his paradigm; he made the observable facts consistent and coherent with the phlogiston conceptual system by his very naming of the substance; it was "dephlogisticated air." Lavoisier, on the other hand, took the stuff and made it the central element in his new theory. He named it "oxygen," and worked up a wholly new conceptual system—called "oxygen theory" naturally—in opposition to phlogiston theory. Priestley never did accept oxygen, or its related theory. He died still believing that what he had isolated was a gas which was totally free of phlogiston, that is, "dephlogisticated air."

Curious Questions

Some very curious questions arise from this sequence of events. One of the most fascinating for me is the problem raised by the query: Did Priestley discover oxygen? In a certain sense I want to say: No, Priestley *could not* have discovered oxygen; he did not even believe in the stuff. It seems to me that what Priestley discovered was dephlogisticated air. And oxygen and dephlogisticated air are two very different things. Phlogiston, whatever it is, is given off during combustion, it combines with calxes to form metals, it most likely has a negative weight, and so on. Dephlogisticated air, therefore, is regular air which has nothing in it with these properties. This stuff is what Priestley discovered, and this stuff is *not* oxygen. Of course, in another sense, Priestley *did* discover oxygen, even though he denied it until his death.

What about Lavoisier? Did he discover oxygen? Again, a difficult problem in both understanding and semantics. It is certain that Lavoisier *created* oxygen, at least in the sense that he invented the concept of a "something in the air which combines with metals during calcination to increase their weight." But when he had first isolated it, he did not realize that it was oxygen—although he *could have* realized it in a way in which Priestley *could not* have realized it. Probably the safest course to steer in this matter is not to make any hard-and-fast declaration about who discovered oxygen, but rather, to let the facts speak for themselves.

A further question is often raised: Was Priestley's discovery an accidental one (no matter what substance it was, exactly, that he *did* discover)? Priestley himself seems to think that his discovery came about because of chance. But I doubt it. Priestley was looking to identify the gas which was given off during the heating of calx of mercury. He knew it had to be *some* kind of a physically typical gas, and it was just a question of finding the test to identify it. The only accident in the whole case was his thrusting a candle into the supposedly exhausted collection bottle. I am very skeptical about whether this event alone is enough to qualify the discovery as being "accidental." A chance occurrence is one element of it, true enough; but the design, the purpose, the intention to identify the gas is always present. And ultimately, it is this intention/purpose/design which motivates even the chance event of the thrust candle.

As a final point, I must return to the main theme of this section. At the start I noted that the identification of an existent object was an important phase of the acceptance of Lavoisier's hypothesis about combustion, calcination, and smelting. Once the object had been captured and some of its properties verified sufficiently for the purposes of identification, confidence in the acceptability of the hypothesis can be seen to have grown measurably. But another and equally important task remained for Lavoisier: He had to use the hypothesis to provide complete internal coherence to the facts of combustion, calcination, and smelting; or if such complete internal coherence were not achievable, then he at least had to provide an internal coherence for the observable facts to a degree equal to or better than that provided by the phlogiston theory. Following provision of internal coherence, the theory can be rendered even more acceptable if Lavoisier can provide for external coherence between the theory and some other subsystem of the overall conceptual core of the tradition.

THE CONCEPTUALISTIC ISSUE

In 1783 in a memoir communicated to the French Academy, Lavoisier attacked the phlogiston theory directly, in terms of its *coherence as a conceptual system.* This memoir, "Reflections on Phlogiston,"[5] explicitly deals with deficiencies in the coherence and consistency of the phlogiston

[5]Quoted in Douglas McKie, *Antoine Lavoisier* (New York: Collier Books, 1962), pp. 110–112. This is a good book to have for information about this case.

theory, and argues that his new oxygen theory provides better coherence than does the old system. I am going to quote a couple of passages from the conclusion of the memoir, just to let you see that conceptual coherence is the main topic of discussion. Then I will briefly describe several series of experiments which Lavoisier ran between 1777 and 1784 in an effort to establish the requisite coherence between experimental observations of oxygen situations. At the end of this analysis it should be clear to you how internal coherence functioned in the acceptance of Lavoisier's new ideas. As a final brief point, I will mention some facets of the external coherence provided between oxygen theory and the physics of Lavoisier's time.

Lavoisier's conclusion first makes a frontal attack upon the coherence of the phlogiston theory. This is an obviously good move: First show that your opponent's conceptual system has an important logical weakness; *then* show that your view does not have this weakness. Here is how Lavoisier leads off:

> All these reflections confirm what I have advanced, what I set out to prove, and what I am going to repeat again. Chemists have made phlogiston a vague principle, which is not strictly defined and which consequently fits all the explanations demanded of it.

Lavoisier points out that phlogiston *does* provide the coherence "demanded of it" in explanation of all the relevant observational phenomena; but at what cost? In order to preserve its coherence, phlogiston has been rendered a vague concept, one which cannot satisfy the strict demands of scientific definition. Lavoisier goes on to back up this assertion, paying special attention to the point that the phlogiston concept not only is vague, but has become *logically inconsistent*—a fundamental and mortal sin against the principle of coherence:

> Sometimes it [phlogiston] has weight; sometimes it has not; sometimes it is free fire, sometimes it is fire combined with earth; sometimes it passes through the pores of vessels, sometimes these are impenetrable to it. It explains at once causticity and non-causticity, transparency and opacity, color and the absence of color. It is a veritable Proteus that changes its form every instant!

Lavoisier makes a rather subtle and delicate two-pronged attack in this passage. First, he shows that phlogiston apparently has contradictory properties, e.g., weight and no weight. That is, phlogiston is "(Wp.~Wp)." Second, not in itself but as an explanatory concept, phlogiston implies—i.e., produces—contradictory properties in the observable substances it is involved with. Thus, phlogiston explains/causes causticity (alkali harshness) and also noncausticity. Accordingly, "If phlogiston, then causticity" and "If phlogiston, then noncausticity." It is a tried and true rule of modern philosophy and logic of science that if a concept explains in the same way both a property and its opposite, then the concept is unacceptable. Lavoisier is clearly relying on this rule for his second criticism. His reasoning is crystal clear and exemplary.

In a passage immediately after those quoted above, Lavoisier makes another significant move. He attacks directly the alleged existence of phlogiston—and his attack is based upon the logic of coherence. Here is what he says:

> My only object in this memoir is to extend the theory of combustion that I announced in 1777; to show that Stahl's phlogiston is imaginary and its existence in the metals, sulphur, phosphorus, and all combustible bodies, a baseless supposition, and that all the facts of combustion and calcination are explained in a much simpler and much easier way without phlogiston than with it.

Lavoisier's argument here is easy to discern. His first premise is that:

> P.1 "All the facts of combustion and calcination are explained in oxygen theory, without use of the phlogiston concept."

The second premise is:

> P.2 "The oxygen explanation is simpler (and "much easier," which perhaps means more "efficient," or more "elegant," although I am not sure exactly what it means)."

From this follows his conclusion:

> C. "Phlogiston is imaginary, its existence is a baseless supposition."

It seems quite clear from this that Lavoisier is arguing in the manner I analyze, that is, *from* logical deficiencies in the concept *to* nonexistence of the substance named by the concepts. Not only are his premises in the argument obviously related to the coherence/conceptualist issue, but also they bring in an admixture of the aesthetician view. In his reasoning Lavoisier concerns himself with the simplicity of the logic, and even though his term "easiest" does not convey a specific meaning to modern minds, we do get from it an aesthetic flavor, a certain value-laden aura which is undeniable. However, he refers here to his theory of 1777 and apparently has already in the preceding sections of the memoir spent some time recounting his work in the interim between 1777 and 1783. What went on during that period?

Internal Coherence

In his experiments between 1777 and 1783, Lavoisier set out to deal with each separate conceptual niche of the phlogiston theory. In turn, he dealt with calcination, combustion, smelting, and ultimately, respiration. Each time he attempted to explain the well-known facts using no reference to phlogiston, but only his concept of "eminently respirable air" (ERA). He also developed new facts, particularly quantitive measurements of a delicate order of accuracy. Needless to say, his attempts were successful.

In 1777 he carried out two series of experiments, the first related to calcination and the second to respiration. Tin calcines fairly easily. A weighed sample is put into an enclosed vessel containing a measured amount of common air. The vessel is heated until the metallic tin has been entirely transformed into an earthy calx. The calx is then weighed and so is the residual gas in the vessel. The calx has gained an amount of mass identical to the amount lost by the common air. Moreover, the common air is no longer common *air*; it will not support combustion or respiration, nor does it pass the goodness test. Lavoisier comes to call this residual air—the remainder from common air after its "eminently respirable" part has been removed—*moffet*, which means "an asphyxiating gas." This experiment allows him to propose the equation given below.
Calcination:

$$\text{Metal} + \text{ERA} \Rightarrow \text{metal calx}$$

According to this description, the calx is a compound of the metal plus the eminently respirable portion of the common air. Note that this is precisely backwards to the phlogiston equation. Moreover, thinking of the ERA as a discrete, independent physical object which can move about during the reaction now further allows Lavoisier to come to a notion about smelting, both with and without the addition of charcoal. Simple smelting, such as that done *without* charcoal as in the case of calx of mercury, can be easily conceived, since it is just the reverse of calcination.
Simple smelting:

$$\text{Metal calx} \Rightarrow \text{metal} + \text{ERA}$$

But in the case of a smelting done with addition of charcoal, the reaction produces fixed air—which Lavoisier now conceives as being a compound produced by movement of the ERA from its location in the calx into some sort of union with the parts of the charcoal. The reaction is fairly straightforward. If

Fixed air = charcoal + ERA

then,
Normal smelting:

Metal calx + .charcoal ⇒ metal + fixed air

As you can see from these equations, all the parts are well accounted for. Moreover, and most importantly for Lavoisier's case, he can also account for all the weights. Indeed, he can use the initial weights of the reactants on the left side of the ⇒ sign to predict the weights of the reactants on the right side. This ability to predict precise amounts is highly significant when compared to phlogiston theory, which can do nothing similar. I must also remark that Lavoisier's use of equations for prediction, as you have probably already figured out, necessarily presupposes the principle of the conservation of matter. His use of this principle is in distinction to many of his chemical contemporaries, and I will note below how it provides evidence of the external coherence of oxygen theory.

Lavoisier's other work, the work on respiration, was reported in May 1777. In this work he started out by assuming Priestley's hypothesis that common air was a mixture of at least two gases, ERA and moffet. Using this analysis, he described respiration as a kind of slow-speed calcination. Thus, if

Common air = ERA + moffet

then,
Respiration:

Body's fuel + common air ⇒ fixed air + moffet

According to this analysis, there is something in animal bodies which acts as fuel, just as does charcoal in combustion. When the common air enters the body, the ERA becomes attached to the fuel, producing the fixed-air compound which is then exhaled. Respiration, according to this conception, uses up the ERA of common air, leaving fixed air and moffet as residuals in the exhalation. To back up this analysis, Lavoisier returns to the calx-of-mercury reaction, and applies the very same notions in a more detailed experimental interpretation. First, he measures the volume of common air which is present in the chamber prior to the calcination. Then he runs the calcination, tests the residual gas, demonstrates that it is moffet, and measures its volume. The equation goes like this:

Mercury + common air ⇒ mercury calx + moffet
(1 volume) (5/6 volume)

On this basis, he concludes that the ratio of ERA to moffet is about 1:5; that is, ERA is about one-sixth of common air. But he does not stop at this point. He goes on to *reverse* the calcination reaction, smelting calx of mercury back to its original state of metallicity. In so doing, he reconstitutes the common air that he originally started with. The equation is in two stages.
Stage 1.

Mercury calx ⇒ mercury + ERA

Stage 2.

ERA + moffet = common air

His analysis of this procedure is fascinating. In the first stage, he dissociates the ERA and metallic mercury; in the second stage, he takes the original 5/6 volume of moffet which remains after the ERA is absorbed during calcination, adds it to the ERA given off during the first stage, and produces the original starting 1 volume of common air. This analysis shows that the ERA hypothesis can be used to render a completely consistent and coherent account of the entirety of facts surrounding calcination, combustion, and respiration. It also accounts for the compound nature of common air and provides an explanation for the evolution of fixed air during both combustion and respiration. There seems to me to be no question about the logical virtues of this account. But of course, it is never easy to disestablish a long-accepted conceptual system. Thus, even at this time—1777—Lavoisier's theory does not have all that many adherents. But he continues his work, and further experiments with the ERA hypothesis add to the coherence among the various observations.

In 1779 Lavoisier coins the name "oxygen" for his new gas. When he does this, ERA takes on an independent life of its own as a specially named object. Later, in 1783, in his memoir to the French Academy, he makes his ultimate statement on the question. I have already quoted from that memoir, "Reflections on Phlogiston," and I wish to finish up with just one more passage from it. By 1783, Lavoisier's conceptual system had been completely tidied up. As we have seen, his concluding arguments in the memoir point to the positive features of his system's logical coherence and consistency, and at the same time, attack the logical inconsistency and lack of simplicity of the phlogiston theory. But Lavoisier is no fool; he realizes full well that getting a hypothesis established is not simply a question of its evidence, coherence, logic, etc. His own words in conclusion to the memoir are interesting in this regard:

> I do not expect that my ideas will be adopted at once; the human mind inclines to one way of thinking and those who have looked at Nature from a certain point of view during a part of their lives adopt new ideas only with difficulty; it is for time, therefore, to confirm or reject the opinions that I have advanced. Meanwhile I see with much satisfaction that young men, who are beginning to study the science without prejudice, and geometers and physicists, who bring fresh minds to bear on chemical facts, no longer believe in phlogiston in the sense that Stahl gave to it and consider the whole of this doctrine as a scaffolding that is more of a hindrance than a help for extending the fabric of chemical science.

Lavoisier notes a distinct feature involved in theory acceptance, a feature I have not previously mentioned. We should probably call his idea a "psychologistic view" since it has to do with one of the foibles of human psychology, namely, "It's difficult to teach old dogs new tricks." There is such a thing as psychological inertia, a kind of mental resistance to being moved into new conceptual areas, and this mental property most likely accounts for much of humanity's initial resistance to all new theories. And science is no different from other human activities. To put it bluntly, scientists are just like other human beings. Why should people change their ideas when the old ideas seem to work well enough? It takes effort to do this, and who wants to expend such effort? Examples of this sort of thing are easy to find. Priestley himself never did adopt Lavoisier's new system. When he gave his reasons, he specifically pointed out that his behavior was an "old dog/new trick" phenomenon, even in the face of the apparent fact that Lavoisier's theory was in every typical way better than the phlogiston system.

Lavoisier brings up, in addition to this psychologistic view, a further and extremely significant point. He refers to "geometers and physicists" who are beginning to adopt the new oxygen theory. I would like briefly to say a few things about this, and also to give a quick mention to one last experiment, since both are intimately related to the external coherence question.

External Coherence

Lavoisier's oxygen hypothesis clearly eliminated any thoughts about the existence of a substance with negative (or, as Lavoisier mentioned above, zero) weight. He had succeeded in weighing oxygen quite accurately, and had traced its mass throughout its reactions. Moreover, his use of the balance had permitted the development of precise quantitative equations. Given these features, there is no doubt why his system appealed to the physicists: Its formal quantitative style, as well as its substantive concepts, were squarely in line with the best physics of the day. But there was even more to his system which appealed to the physicists and mathematicians. His early adoption of the principle of the conservation of matter fitted rather nicely into the numerical schemes of physics, in which various quantities such as momentum (mv), force (ma), and kinetic energy (mv^2) were all *conserved* entities. The numerical equations that he developed provided strong evidence of conservational laws, which was a new aspect of chemistry. They provided a strong motivation to physicists. Indeed, Emile Meyerson's masterful analysis in *Identity and Reality* argues vigorously that this feature of Lavoisier's program was a telling blow against the nonconservative entities which functioned in the phlogiston system, nonconservative entities which physicists would find somewhat repugnant.[6] Thus, all these features—positive mass, quantitative formalism, precise numerical prediction, and conservational entities—were very attractive to physicists and other mathematical scientists. Thus, what we must see here is a growing external coherence between the new chemistry and the prevailing physics. We cannot discount this aspect of Lavoisier's hypothesis; certainly it had something to do with the rapid acceptance of the chemical hypothesis in neighboring scientific concept structures such as physics. It is also noteworthy that Lavoisier himself specifically mentioned the attraction of his system for physicists.

One of the "geometers" Lavoisier mentioned was the mathematician Laplace, who eventually worked with Lavoisier. Together these two men did experiments which built further bridges between the new chemistry and the older, well-established physical paradigms. One in particular is so brilliant, so astounding in the cleverness of its reasoning, that I must bring it up as a final example of the growing external coherence provided by the new hypothesis. It concerns respiration and heat.

Lavoisier and Laplace came to measure heat by the amount of ice which could be melted by the hot body. Although this procedure provides no absolute measure of heat, it does give a clean, clear, relative value which can be used to compare two or more bodies. Thus, two bodies which each melt two cubes of ice have the same amount of heat; one body which melts only one cube has only half the heat of either of the

[6]Emile Meyerson, *Identity and Reality* (New York: Dover Publications, Inc., 1962), pp. 168ff.

first; and so on. In their experiment using this method, the two men set up a guinea pig in a chamber, and measured how much ice the animal could melt with his body heat over a measured period of time. They also collected the fixed air respired by the animal, and measured its volume. Now recall Lavoisier's hypothesis about respiration: Animal bodies have a charcoallike fuel, which is slowly combusted with the oxygen they breathe in; fixed air is exhaled as a result of this reaction. Thus, the amount of fixed air respired is directly related to the amount of fuel which is burned in the animal's body.

Given these aspects of his theory, Lavoisier now makes a bold hypothesis: The heat given off during the slow-speed combustion in animal respiration should be closely related to the amount of heat which could be generated by burning an identical weight of charcoal. This new prediction follows strictly logically from Lavoisier's concepts; but it is a completely new notion as far as the physics of his time is concerned. The obvious problem, however, is to figure out how much fuel the animal burned during the time period. But Lavoisier comes up with a brilliant method. He has measured already the volume of exhaled fixed air. So what he does now is to burn enough charcoal to produce an amount of fixed air identical in volume to that exhaled by the pig. Then he measures how much charcoal was burned in order to produce that volume of fixed air. Finally, he takes an identical amount of charcoal, burns it, and measures how much ice *it* melts. *Voilá!* The amount of heat produced by the burning charcoal is exactly identical to the amount produced by the pig during its respiration. The reasoning here is quite delicate, moving from heat to fixed air to charcoal and back to heat. With this series of movements Lavoisier completes the circle and, in so doing, makes a firm link bridging pure chemical concepts such as "fixed air," "oxygen," etc., and the physical concept of "heat," not to mention the biochemistry and physiology involved in respiration. Large amounts of external coherence are gained in the move.

By this point, 1784, phlogiston theory was completely doomed; oxygen theory was assured the ascendant position. Lavoisier, by his own statement in 1791, confirmed that the revolution in chemistry was over and done with. All chemists had been forced to come to grips with Lavoisier's new system of concepts; additionally, no few physicists and "geometers" had been similarly exposed. The result was the general acceptance of Lavoisier's hypothesis.

CONCLUSION

The factors involved in this case of theory acceptance are not easy to sort out. I have tended to emphasize the related aspects involved in the metaphysical/existential question and the conceptualistic question, but this has been done only because I happen to believe that these were the two *major* reasons underlying Lavoisier's eventual triumph. Needless to say, however, there are alternate accounts which might be given. A predictor, for example, could very well point to the fact that Lavoisier's later theory allowed very accurate predictions, ones that were in general verified by experiment and observation. An aesthetician could just as well point to the simplicity and elegance of Lavoisier's hypothesis—a point even mentioned explicitly by Lavoisier himself—as telling blows. And so on and so on. I could not agree more that the situation of theory acceptance in general, let alone in this case, is extremely complicated, and that it involves all the factors I discussed earlier as well as others which I have not mentioned. Even so, I would have to argue that the metaphysicians and conceptualists have laid out the most important principles for us to use in the interpretation of the acceptance of oxygen theory. Each of these two views is crucial in the understanding of a significant phase of the processes following Lavoisier's initial 1772 postulation of the mysterious "something" from the air. I might also remark, as a very last point, that the logicians also bring some of their views to bear here, notably in the sense that conceptualism—in either type of coherence—ultimately rests upon a firm foundation in the laws of logic: consistency, deductions, predictions, and so on.

SUGGESTIONS FOR FURTHER READING

In addition to the suggestions found at the end of Chapter 5, there is another useful book which provides some excellent general background information about the men and ideas of the time of Lavoisier. This is Bernard Jaffe, *Crucibles* (Greenwich, Ct.: Premier Books, 1964), especially chaps. III, IV, V.

Chapter 10

Acceptance of Pasteur's Hypothesis

INTRODUCTION

In Chapter 6 I left you with the thought that Pasteur's general hypothesis, his fundamental discovery, was the proposition "All fermentations require living beings (yeasts)." But I also noted that Pasteur's evidence for this generalization was extremely weak, since it involved at least two other specific hypotheses which had not yet been accepted, "Alcoholic fermentations are the necessary results of the activities of living beings," and second, "Lactic acid reactions are fermentations which require yeasts." Proving any of these involves proving all of them, and almost at the same time. Moreover, as I also noted, Pasteur's hypotheses here are in direct opposition to the paradigm conceptual system which prevailed at the time. Thus, getting his discovery accepted is doubly hard, not only because it is controversial, but also because the evidence is both extremely weak and quite circular. But Pasteur's arguments very quickly won the day; his theory was rapidly accepted, and soon became not only noncontroversial, but indeed the core paradigm for a rapidly expanding independent tradition we now call "biology."

My analysis of the events leading to acceptance of Pasteur's hypothesis will focus most closely upon two aspects of the process. In the first place, Pasteur's arguments, like Lavoisier's, must necessarily involve reference to an existential/metaphysical issue. Pasteur's hypothesis makes the claim that microorganisms are responsible for fermentations in general, and lactic acid fermentation in particular. There is no doubt that brewer's yeast is an object and, moreover, that it is an object often associated with alcoholic fermentations. However, and this is crucial, there has been no apparent connection between any object and the lactic acid reaction. If his hypothesis is even to gain initial plausibility, Pasteur must come up with such an object. This he does. But even this is just the first step in the metaphysical problem; he also must show that this object is the *cause* of the reaction, i.e., that its activities are necessarily related to the lactic acid production. This he does, as I will show, by use of a technique drawn from an especially clever analogy between "yeasts" or "ferments," and "seeds" which are sown in the ground. I call this extended argument the "farming analogy." I will go on to argue that the later history of the success of Pasteur's hypothesis is in large part the story of the successful deployment of this analogy. But Pasteur's first job is to use his "sowing" technique. Once this is done, it is plausible that lactic acid reaction is a fermentation, with an associated yeast. Moreover, *this* plausibility adds to the plausibility of the view that "Brewer's yeast is necessarily associated with (causes) alcoholic fermentations." Both plausibilities in conjunction add to the final plausibility of "All fermentations require living beings." Thus, finding an object, a new "yeast," and bringing it back alive from the jungles of the lactic acid reaction, will be an extremely solid blow in favor of the acceptance of Pasteur's hypothesis.

The existence of such an object, however, is not the only main feature involved in the eventual acceptance of Pasteur's theory. There is another, a feature strongly related to the conceptualistic issue of coherence, but one we cannot properly call just plain "conceptualistic." Because of this difference, I will now amend conceptualism vis-à-vis theory acceptance. As earlier discussed, the conceptualistic criterion generally involves the growing external and/or internal coherence provided by the hypothesis for known (and new) observational facts. Whenever coherence grows, acceptability increases. But this version of general conceptualism focuses entirely upon the *intellectual* aspects of the concepts; it does not say much at all about their practical ramifications. This is a deficiency in the conceptualistic view, because in the Pasteur case, I must claim that the *practical consequences* (or at least the *projected* practical consequences) of growing coherence, and not the growing coherence alone, were decisive in the quick acceptance of his theory. In order to adapt conceptualism to this fact, the conceptualistic notion about theory acceptance must be modified, divided into two stages. Theory acceptability grows as: (1) the new hypothesis provides a growing conceptual unity within and/or between different sets of notions; (2) the new unities provide significant projected advances in practical affairs. Let me give you only one very brief example of the sort of thing that I mean. My rendition will also show you the link between the practical aspect of conceptualism I am discussing and a certain interpretation of the predictor position.

Pasteur's hypothesis predicts that beer spoilage is simply a fermentation other than the desired alcoholic one. Moreover, it further implies that the cause of the spoilage fermentation is a living being, a different sort of "yeast." Finally, it follows that, if this being can be killed or otherwise controlled, and the beer rendered isolated from any further infection, then the spoilage will be prevented. The potential or projected practical consequences of this set of inferences are obvious; for example, it is well known from normal, everyday observations that high heat kills living organisms. But given the new hypothesis, this everyday observation takes on a new consequence. That is, one can make this everyday observation *practically* cohere with the new hypothesis by developing a high-heat process for beer after its production but before its spoilage. This heating process of course *was* developed and, again of course, came to be called "pasteurization" for obvious reasons.

What I am getting at here is the clear fact that Pasteur's core hypothesis had an immediate set—a vast and diverse set—of potentially valuable practical consequences. Thus, if the view were accepted to be true, then it would mean that a whole set of practical problems could possibly be dealt with. The interesting point of this is that it was not necessary *first* to achieve the practical effects and *then* to accept the view; what actually happened was that his contemporaries merely *perceived* the inestimable potential of the practical effects, and on that basis accepted the theory. This rather odd procedure, namely, accepting hypothesis on its perceived or projected practical potential, rather than on its achieved practical actuality, is why I must insist that this reason for acceptance of Pasteur's hypothesis is not purely a *pragmatic* one, but rather is more properly categorized somewhere under conceptualistic views. I will call this idea of mine "conceived practical richness" in order to indicate that it mixes conceptual and pragmatic reasons, but that the conceptual reasons predominate, at least during the period of acceptance itself.

THE METAPHYSICAL ISSUE

It would probably help us all if I reviewed the whole situation confronting Pasteur. We are located in about 1859. Pasteur has done his work for the brewers of Lille, and he has also come up with his hypothesis about the mechanics of the lactic acid reaction. His work has been announced, and he is continuing to follow out the implications of his hypothesis "All fermentations necessarily involve living beings."

Opposed to his view is another conceptual system, the chemical paradigm best represented by Justus Liebig, well-respected German chemist.[1] According to Liebig, fermentation is a purely chemical, i.e., mechanical, nonliving reaction. It involves two ingredients: (1) a fermentable substance such as beef broth, grape juice, or in the most typical case, sugar water; (2) a nitrogenous substance which acts as the chemical agent of the reaction, yeast being a typical example. According to this view, the dynamics of the reaction go like this: The nitrogenous stuff (n-stuff) is dead and the fermentable substance has a peculiar sort of instability. When the n-stuff starts to decay or decompose, it triggers the instability in the fermentable substance, which causes it also to decay or decompose. During the decomposition CO_2 (fixed air) is given off, the reaction foams vigorously, and debris—mostly more n-stuff—collects on the bottom of the vessel.

There is a whole group of reactions which are conceived to be fermentations according to this definition. The alcoholic fermentation of beer and wine is a well-known instance. Other examples include reactions which produce lactic acid (when milk sours) and butyric acid (when butter spoils). Among all these various fermentings, only the alcoholic one is known to have a living agent (the brewer's yeast) associated with it. However, the fact that yeast is alive has absolutely nothing at all to do with the alcoholic fermentation. Rather, it is the dead material, the decaying n-stuff, among the yeast colony which triggers the alcoholic reaction.

Pasteur's hypothesis denies this prevailing chemical paradigm at all possible points. First, he claims that the life processes of brewer's yeast are essential to the alcoholic reaction. Second, he asserts that, although none has ever been found, each and every fermentation must have its own individual associated living system. In particular, lactic acid fermentation necessarily has its own special new "yeast" or microorganism functioning as the causative agent in the reaction. One problem remains, given these first two points. According to the dominant view, the n-stuff is necessary to the reaction since it is the causative agent. But this Pasteur denies. He claims instead that a living system is the cause. In this case, then, what is the role of the n-stuff? All fermentations do indeed demand it; but why? Pasteur must answer this question, which he does in a straightforward way. If, he says, the causal agent is a living system, then it needs food.

[1]James B. Conant (ed.), *Pasteur's Study of Fermentation* (Cambridge: Harvard University Press, 1952), Case 6, p. 33.

The role of the n-stuff is to be food for the microorganism as it works through the fermentation. Later on, Pasteur slightly modifies this analysis into terms relevant to the farming analogy. N-stuff, on this later view, functions in the same way as does manure applied to a seeded field.

It is extremely interesting to see how Pasteur's conception of this whole situation changes over a short period of time. My account above has been taken from two papers, one, which you already know of, prepared in 1857; the other, which you have not yet heard about, read in 1860 ("Memoir on the Organized Corpuscles Which Exist in the Atmosphere"). In this second memoir, Pasteur has become quite clear on the general issues which separate him from the rest of his tradition. By this time he has identified several more new microorganisms and used them to carry out their respective fermentations. He is more confident, and he both sees and states the issues more sharply. But in the earlier paper he is also quite confident, although only about the more restricted issue of the lactic acid fermentation in particular. Let me quote from the introduction to that earlier paper, in an effort to show Pasteur's clear and explicit recognition of the metaphysical question involved in his new hypothesis. He states:

> In the first part of this work, I plan to show that just as an alcoholic ferment exists, namely, brewer's yeast, which is found wherever sugar breaks down into alcohol and carbonic acid, so too there is a special ferment, a lactic yeast, always present when sugar becomes lactic acid, and that if any nitrogeneous plastic material can transform sugar into this acid it is because it is a food suitable to the development of this ferment.[2]

This passage is fascinating because of the collision between what Pasteur tries to do in its first few clauses and what he in fact claims in the last couple of clauses. In the first bit, he does not really explicitly say that the brewer's yeast *causes* alcoholic fermentation—which he actually believes. Rather, he edges up on the proponents of the chemical paradigm, using terms such as "found whenever" and "ferment," which are neutral as between his own view and that of the chemists. Both viewpoints can use the term "ferment," although each means something different. Moreover, the lactic yeast "is present" only; he does not explicitly attack the causal account given by the chemical view. His sole area of attack is directed to the point that a lactic yeast exists, a view which had been denied by everyone else prior to him.

The last section of the passage, however, is a direct frontal assault upon the prevailing view. Pasteur here uses causal language ("trans-

[2] Ibid., p. 28.

form"), with which he denies specifically the chemical view that the n-stuff—"nitrogeneous plastic (decomposable) material"—is the agent of fermentations. The n-stuff simply is not the cause of fermentation. Rather, he asserts, the n-stuff functions as a food for the reaction and not the causal agent.

Even though three elements are involved in his stated objectives, Pasteur, on the one hand, quietly assumes the first in a cleverly ambigious fashion, and on the other, never, in the paper, offers direct proof of the third. Only the second claim, that lactic acid fermentation has a causal microorganism, is explicitly demonstrated in the work of the paper. But this, of course, is a vast enough task to accomplish in one fell swoop—even for a giant such as Pasteur.

Although I might seem to be stretching things just a bit, I feel that I must claim that Pasteur came as close to settling an existence question in one paper as anyone ever has. He just could not have come any closer in one shot to resolving the question whether or not an object required by a hypothesis exists. Nicely enough, at least in terms of the account provided by Harré's analysis of existence questions, Pasteur settles the metaphysical/existence issue by developing a brilliant analogy between his new concept and an already well-known area, namely, farming.[3] Then on the basis of the analogy, he develops a revolutionary new technique which single-handedly opens a whole new area of researches.

The Farming Analogy

It had been known for some time that microorganisms were living beings. Microscopic researches had discovered various protozoans—single-celled animals such as amoebae and paramecia—and plantlike creatures such as euglenae. Then, about ten years prior to Pasteur's work, a new, powerful type of microscope had come into use, extending far lower the limits on perceivable small beasts. This new machine had fouled up the taxonomic system, since it brought into view new objects such as bacteria, which were not easily classifiable as either plant or animal. However, even though it caused problems for taxonomy, the powerful new microscope aided microbiology enormously. It is clear that Pasteur's finding of the new lactic "yeast" was made possible only by the increased range of the new instrument.

Yeasts themselves had been known for quite a while to be living, breathing, breeding creatures. They were generally thought to be species of plant, with all the features attendent upon plant ecology—excepting, of course, those activities, such as photosynthesis, which are related to

[3]Rom Harré, *The Philosophies of Science* (Oxford: Oxford University Press, 1972), p. 170ff.

green pigments (chlorophylls) which yeasts do not have. Otherwise, yeasts had been observed budding, creating daughter cells, moving around, and in general, living life as they best knew how. So there was no question at the time about the aliveness of yeasts, nor of their plantlike nature. What was in doubt (indeed, "doubt" is too weak: what was not believed) was that there was any connection between the life processes of yeast and the fermentation reaction itself. Pasteur changed all this. He took the accepted notion that yeasts were plants, and then developed it vastly into a full-blown model of what microagriculture might be like.

He started out by comparing yeast to plant seed. This comparison alone is very conceptually fruitful. For example, everyone knows that each specific type of seed produces a specific type of crop. Thus, if you want to grow watermelons, you do not plant rutabaga or turnip seeds. By analogy, it follows that the alcoholic yeast is the *alcoholic* yeast, and not the lactic acid yeast. Generalized, this specificity rule requires that each fermentation have its own particular special seed or "yeast." Moreover, on this model, the fermentation reaction itself is actually the crop-growth cycle. This part of the analogy is very nice. It explains, for example, why there is more yeast *after* the beer fermentation than there was before. Prior to Pasteur's use of the farming analogy, there was no real explanation (that is, no explanation that was not ad hoc) for why, when it was supposed to be decomposing, the yeast colony was rapidly, and enormously, increasing in size. To put the case in an odd but revealing way, the farming model implies that brewing beer is actually the same thing as growing a crop of brewer's yeast.

Further implications of the model come to mind immediately. Apparently, the sugar water, grape juice, or other medium is the same thing as a well-prepared field, ready for planting. The medium, within broad limits, can grow anything that happens to fall into it, just as a nicely tilled plot can grow—again within limits—corn just as well as other varieties of plants. For another point, take the problem of weeds. Every gardener knows that vegetable (or other) plots are always getting messed up by various noxious things we are not interested in—bindweed, dandelions, Johnson grass, you name it. According to Pasteur's analogy, we can expect exactly the same sort of thing to happen whenever we get a microfield ready for planting a microcrop. Spoilage organisms are the weeds of the microecology.

The weed notion answers another plaguing question. It had always been asked, "Where do the spoilages come from in the first place?" When Pasteur formulated the farming analogy, the answer suddenly appeared, as if it had been there all along. You might as well ask, he noted, "Where do weeds come from?" Weed seeds are in the air, everywhere. They are ubiquitous, universal, inescapable, a necessary and essential part of our gardening and lawn-growing life. The exact same thing holds, according to the farming analogy, for the molds, bacteria, fungi, yeasts, and other organisms of the microworld. They are in the air everywhere, just looking for a hospitable "plot of land" such as a glass of warm milk, a pat of butter, a fresh-fallen ripe fruit, a leftover lamb chop. Put the seed together with the plot, and a crop results. Pasteur developed and rapidly deployed these analogical notions. Already by the time of the 1860 paper he was well into exploiting the consequences of the farming model. We need only remark the title of the 1860 work: "Memoir on the Organized Corpuscles Which Exist in the Atmosphere."[4] The "organized corpuscles" have already become wind- and airborne "weeds."

Seeding the Plot

From this model, Pasteur was led directly to a revolutionary new technique. He reasoned that if the microorganisms were similar to crop seeds, then they could be sown into a prepared "plot." He proceeded to do just this. First, of course, he had to get the seeds themselves. In his first attempt at this technique, he looked carefully through all the regions of the lactic acid fermentation vessels, and began to notice a very obscure sort of "globule," as he calls it. It was only about half the size of a regulation brewer's yeast, but it was otherwise relevantly similar. It was protein-containing, somewhat transparent, and obviously organized in its structure. Moreover, it universally accompanied lactic acid fermentations. Pasteur leapt, with good reason, to the conclusion that this tiny globule was the new kind of yeast demanded by his farming analogy. His next moves were obvious. He isolated the beast, and put it into a specially prepared nutrient solution: sugar water, boiled yeast extract (for n-stuff), and a little chalk to keep the alkalinity correct (you may know that farmers often chalk, or lime, their soils to keep the earth sweet). What happened was no surprise to Pasteur, even though it rather raised his colleagues' eyebrows. The nutrient solution went into a clean, hard, fast lactic acid reaction. So he did it again, and the same thing happened. Further repetitions produced the same results, which were plain enough for anyone to see. The compelling features of his demonstration were dramatic increases in the regularity, purity, and strength of the reaction. It had become fairly routine by then to produce lactic acid via exposure of the appropriate medium to the air. But these "wild"—as opposed to "domesticated"—reactions were not entirely surefire. Pasteur himself noted that oftentimes an alcoholic reaction would develop concurrently

[4]James B. Conant (ed.), *Pasteur's and Tyndall's Study of Spontaneous Generation* (Cambridge: Harvard University Press, 1953), Case 7, p. 16.

with the lactic reaction. Sometimes, indeed, the medium would refuse to grow any lactic acid at all. But Pasteur's hand-sown crops developed uniformly and with extremely high regularity. Moreover, the reaction was quite clearly more rapid in both onset time and length of "season" required to produce the crop. Finally, the crop was pure lactic acid, of a concentration beyond anything typically achievable.

Although there could not be many doubts remaining that Pasteur was "growing" the lactic acid organism, it would be far too much to expect that the evidence from this one reaction provided conclusive proof for his contemporaries. Pasteur himself said as much in the conclusion of the paper:

> All through this memoir, I have reasoned on the basis of the hypothesis that the new yeast is organized, that it is a living organism, and that its chemical action on sugar corresponds to its development and organization. If someone were to tell me that in these conclusions I am going beyond that which the facts prove, I would answer that this is quite true, in the sense that the stand I am taking is in a framework of ideas that in rigorous terms cannot be irrefutably demonstrated.[5]

Even though his ideas *were* bold, and opposed to the prevailing view, Pasteur went on to ask a fair hearing for his ideas, and to restate precisely his general hypothesis about the relation between life and fermentation, a general hypothesis for which the lactic acid reaction provided only one piece of data:

> . . . Whoever judges impartially the results of this work and that which I shall shortly publish will recognize with me that fermentation appears to be correlative to life and to the organization of globules, and not to their death or putrefaction.[6]

In this passage Pasteur specifically refers to and denies the theory of Liebig and the chemists, namely, that the fermentation reaction is a decomposition produced by the decay/decomposition of the n-stuff contained in globules.

But Pasteur's minority position was not long sustained. Within two years of the above memoir Pasteur's attitude apparently had changed to one of confidence that his ideas had already become accepted. A short paper, a note actually, was given to the French Academy on June 17, 1861.

[5]Conant, Case 6, op. cit., p. 32.
[6]Ibid.

Its title was "New Experiences and Insights on the Nature of Fermentations." In it, he first stated his now familiar general hypothesis:

> . . . All my efforts were directed toward demonstrating that fermentations were correlative to the presence and proliferation of living organisms, with a different organism corresponding to each type of fermentation. . . .[7]

He then went on to claim that, as regards this general leading idea, "any possible doubts which may have remained in a few people's minds must have been removed by the results which I recently had the honor of presenting before the Academy, on the subject of butyric fermentation."[8] In the memoir referred to, Pasteur had successfully used his new "farming" technique in order to hand-sow a crop of butyric acid bacteria. With the success of this additional demonstration of his theory, a demonstration carried out in a very different type of reaction than the lactic acid fermentation, Pasteur believed that his hypothesis could now only be accepted. As he says, in spite of initial opposition, "I make bold to hope that it can be regarded today as an addition to scientific knowledge."[9] Hypothesis no longer, but knowledge.

The Final Discovery: Fermentation—Life Without Air

There remained one serious deficiency in Pasteur's theory, one point which remained to be cleared up relative to the conceptual system of his opposition, and it is a point which will figure nicely in the discussion of Pasteur and Lister, below. According to the prevalent theory, spoilage and fermentation were necessarily dependent upon oxygen. Canning, the method of food preservation we are all familiar with today, had been discovered not long before Pasteur's work. Although agreement was not universal about how canning worked, its mechanism was usually explained by reference to the fact that there remained no free oxygen in the vessel containing the canned preserve. Since there was no oxygen, there could be no spoilage. On the other hand, Pasteur's theory, which required the causal agent of fermentation to be *alive*—and life requires oxygen—obviously had somehow to deal with the question of the oxygen mechanism of the living ferments. This question was answered for good in the 1861 memoir we are now discussing. Moreover, Pasteur managed, in his explanation, to hang on to the term "decomposition," which was of benefit in making his theory more acceptable to the chemists. Let me now quote for you the passage which contains Pasteur's hypothesis about the role of oxygen in the life processes of yeast:

[7]Hilaire Cuny, *Louis Pasteur* (New York: Paul S. Erickson, Inc., 1966), p. 165.
[8]Ibid.
[9]Ibid.

> In this lies the whole mystery of fermentation! For if the question I have asked be answered by saying, "Since brewer's yeast absorbs oxygen plentifully when the gas is present in the free state, the reason must be that it needs oxygen in order to live, so that it has to take oxygen from the fermentable medium when no free oxygen is present"—the plant is immediately revealed to be an agent of the decomposition of sugar: at every respiratory movement on the part of its cells, there will be molecules of sugar whose equilibrium is destroyed by subtraction of part of their oxygen.[10]

With a masterful leap of reasoning, Pasteur here solves the mystery of fermentation and, at the same time, plugs a serious gap in his own theory. Yeasts, since they are alive, need oxygen to live. But where do they get this oxygen? In some circumstances, they simply pick it up from among the dissolved gasses in their medium. But in brewing circumstances, there is a blanket of CO_2, which is heavier than air, shutting off their medium from replenishment of oxygen. What then? Pasteur here theorizes that yeasts have a second mode of life: When they are not in the presence of free oxygen, they manufacture it from one or more of the constituents of their environment, chiefly sugar. In the process of "mining" oxygen from the sugar, the yeasts incidentally decompose that sugar into alcohol and CO_2. Thus, the mechanics and dynamics of the fermentation reaction are plain and clear. As Pasteur points out, "we are led to wonder whether there may not be a hidden relationship between the property of being a ferment and that of living without atmospheric air."[11] This is the immediate precurser of his soon-to-be-announced definition of fermentation as "life without air."

By this time, the game was over. Pasteur's theory had won the day, within four years of its first announcement. This is not to say that there were no objectors in the ensuing years. That is too much to be expected, especially in view of the fact that I have defined science itself in terms of the trial-by-fire processes of discovery and acceptance. The objections, interestingly enough, were somewhat nationalized in their origins; most of the objectors were German. For example, Pasteur, in his 1879 definitive work on fermentations, takes one whole section of the work to respond to objections raised by the German scientists Brefeld and Traube. Brefeld's memoir of 1873 is quoted specifically. Brefeld notes that Pasteur's whole theory commands "general assent," but that he, Brefeld, has done experiments which show it to be "untenable."[12] Brefeld's work, which we need not go into here, apparently showed that ferments could not live

[10]Cuny, op. cit., p. 168.

[11]Ibid., p. 166.

[12]Louis Pasteur, "The Physiological Theory of Fermentation," in Charles W. Eliot (ed.), *The Harvard Classics*, vol. 38 (New York: P. F. Collier & Son, 1910), p. 329.

("increase") without free oxygen. Pasteur of course attempts to rebut Brefeld's claims. The point I wish to make here is that, even among his opponents, Pasteur's theory had become accepted, and indeed it had become the established theory, by 1873.

The role of the metaphysical issue is obviously of extreme significance in this case. Pasteur's hypothesis was intimately and inextricably bound up with the existence of an object which caused certain changes in fermentable substances. His whole objective then was to find such an object, and to demonstrate that it had the powers and properties needed to carry out its hypothesized tasks. His revolutionary technique of seeding a microorganism also was of major effect in the acceptance process. The seeding technique made it extremely plausible that the microorganism was responsible for the fermentation since it was possible to isolate and then follow observationally the behavior of the purported fermentational agent.

But the existential issue is not the only issue. Also working in Pasteur's favor was the coherence which his hypothesis provided. Moreover, the coherence provided as well a growing sense of the potentially significant practical ramifications of Pasteur's new concepts.

COHERENCE AND PRACTICALITY

I pointed out earlier that, in one important sense, the acceptance of Pasteur's hypothesis that fermentations were intimately correlated to the life processes of living beings was also, at the same time, the acceptance of his farming model of microorganisms. I will now spell out this claim a bit further, in hopes of showing you how accurately it describes the case.

The essential points of the farming analogy were plain for everyone to see. I must emphasize that when Pasteur argued for his hypothesis "There exists a lactic acid agent," he was not simply arguing for a restricted proposition. That is, given the total logical structure in which his argument was arrayed—the analogy itself—his audience was not being asked to accept this proposition alone, but rather the whole vast conceptual system bound up in the agricultural model. Thus, in effect, if they bought the lactic acid hypothesis, they were buying as well the whole new way of thinking about microorganisms and not just a modest proposal about an obscure bacterium unobtrusively making lactic acid in a quiet corner of the dairy. But the very scope of the model, its powerful and wide-ranging implications, seems to me to have been an enormous asset, and not a liability. Had Pasteur been asking his colleagues merely to accept the restricted hypothesis, it would not have portended any new world of concepts and practical effects. But his proposing an hypothesis which was embedded in a whole new way of looking at things—his offering to them of a whole new world of agriculture—forced immediate and careful scrutiny of his ideas. These ideas, as was immediately apparent to all, might very well contain something *important.*

I must here mention something which I have been studiously avoiding up until now. This is the medical issue. I started off my discussion of Pasteur in Chapter 6 with the declaration that Pasteur's significance was located, not in his medical advances, but in his theoretical breakthrough in microbiology. I still stick with that belief. But I have to let you in on a couple of things which happened during the acceptance phase of his microbiological paradigm.

A connection, at least a conceptual link, between disease and fermentation had long and strongly been suspected. Robert Boyle, a seventeenth-century chemist we have met before, stated his own views on the subject explicitly and succinctly:

> And let me add that he that thoroughly understands the nature of ferments and fermentations shall probably be much better able than he that ignores them, to give a fair account of diverse phenomena of several diseases (as well fevers as others), which will perhaps never properly be understood without an insight in the doctrine of fermentations.[13]

Descartes, a French contemporary of Boyle, also remarked the obvious parallels between such things as compost heaps, fermentations, and diseases. Later, during Pasteur's time, Liebig himself believed that the cause of all fermentations and most contagious diseases was the same, namely, the destabilizing action of a nitrogeneous ferment. This idea was held by all parties, chemists and antichemists, at the level of received dogma. Scientists in all fields accepted it, did not question it, and carried on their work in terms of the supposed connection. I think you can already see the incredible implications of Pasteur's new conceptual system when it is linked up with this widespread belief in a connection between fermentation and disease. For instance, look what so quickly happens when you first deploy the farming model into the new terrain constituted by medical affairs. One of the first things cranked out is the idea that disease, like fermentation, is the growing of some crop of "plants." Moreover, the specificity rule for "crops" now applies to diseases as well as to fermentations: Each specific disease has its own specific causal agent. And each of these agents has its very own life-style:

[13]Robert Boyle, *Essay on the Pathological Part of Physik,* quoted in Conant, Case 6, op. cit., p. 50.

Some like it hot, some like it cool; some like it wet, some like it dry; and so on. Once this conceptual light has gone on, ideas about possible preventions and cures can be generated almost automatically.

Lister's Hypothesis

The very best example of the conceptual fruitfulness of Pasteur's hypothesis occurred almost immediately in the work of Dr. Joseph Lister in Glasgow, Scotland. At the start, Lister believed in the chemical theory of fermentation/disease; in particular, he held to the necessary role of oxygen in these reactions. But when he read of Pasteur's researches, he saw immediately the possibilities offered by the new system of thought. He immediately set out to test the implications he had developed. As far as my argument here is concerned, I could stop the Lister story at this point, without even mentioning whether or not Lister's experimental tests were successful. Remember what my point is in regard to the conceptualistic issue: The main thing in the acceptance of Pasteur's theory is the *potential* practical effect of the growing coherence, not its *actual* practical success. That is, if the theory *promises* to have practical effects in a growing number of areas, then its degree of acceptance grows in direct proportion. Of course, it does not hurt at all if some immediate practical success is achieved. But such success, while sufficient for acceptance, is certainly not necessary to it. In any case, Lister *did* succeed; his success made fast fame for Pasteur's ideas. But let me give you Lister's own words, from the beginning of his famous paper, "On the Antiseptic Principle of the Practice of Surgery" (1867).[14] Lister starts out giving his version of the chemical theory of infection:

> In the course of an extended investigation into the nature of inflammation, and the healthy and morbid conditions of the blood in relation to it, I arrived several years ago at the conclusion that the essential cause of suppuration [infection[15]] in wounds is decomposition brought about by the influence of the atmosphere upon blood or serum retained within them. . . .[16]

[14]Joseph Lister, "On the Antiseptic Principle of the Practice of Surgery," in Eliot, op. cit., p. 271.

[15]I am very hesitant to use the word "infection." Lister does not use it, at least not at this early date. An "infection" is something, a process, which necessarily presupposes an agent doing the infecting. It seems to me quite clear that the term "infection" can come into legitimate use only *after* the living-agent model of disease is accepted. This is perhaps the best explanation of why Lister himself, in this early paper, does not use the word which we, in the present time, take completely for granted. Lister uses the term "suppuration," which means "discharge of pus"; we would automatically say "infection." This seems to be an excellent example of how acceptance of a new theory causes linguistic change; moreover, as acceptance grows, the new conceptual system gradually filters down from the purely scientific realm into the world of ordinary mortals. Hence, "suppurate" becomes "infect."

[16]Ibid.

There are a couple of things to note here. In the first place, Lister advances a theory we are by now entirely familiar with. Disease, just like fermentation, is a "decomposition" reaction in which tissues surrounding wounds inflame, and then decay and decompose. The tissues, needless to say, are nitrogenous proteins, and thus are ready-made to support the kinds of "fermentations" or "spoilages" we have seen so often before. Second, Lister refers to the atmosphere as the cause ("influence") of the decomposition. In particular, "oxygen . . . was universally regarded as the agent by which putrefaction was effected."[17] This theory we have also seen before. As in the canning example, oxygen is the causative agent—at least in the sense that the n-stuff, the ferment, could not decompose unless it had a free oxygen supply. In sum, then, Lister's theory is that wounds have an n-stuff, strictly comparable to the n-stuff of ferments, which, in the presence of oxygen, spontaneously decomposes, with the result that the tissue is infected.

Lister's next statement is a marvelous example of how a conceptual system controls the sorts of experiments and observations which come to be attempted. Given his belief that the oxygen of the atmosphere plays a necessary and essential role in suppuration and decay of wounds, Lister at first simply could not even conceive anything practical to do to prevent the tissue decomposition. He says this clearly and succinctly, and in a way which reveals to us the pain its admission causes his sentiments as a physician who wants to alleviate human suffering:

> To prevent the occurrence of suppuration with all its attendant risks was an object manifestly desirable, but till lately *apparently unattainable,* since it *seemed hopeless* to attempt to exclude the oxygen. . . .[18]

Lister's frustration here is apparent. But his frustration reveals in addition a point of philosophical significance; "apparently unattainable," he says, and it even "seemed hopeless to attempt" to prevent suppuration. Here we see the ultimately frustrating epistemological problem disclosed by a realistic interpretation of the role of hypotheses in science. Lister's conceptual system gives him a coherent explanation of suppuration; oxygen plus a disease "ferment" are the causative factors. But on this hypothesis it is hopeless even to *attempt* to prevent the decay. So the experiment simply does not get done. In fact, it cannot even be *conceived* to be practical to try to exclude oxygen from a wound; oxygen exclusion holds zero promise, and thus no observations, no experiments, are even contemplated. In this case we see clearly the role of the hypothesis: The

[17]Ibid.
[18]Ibid. Emphasis added.

hypothesis rules, it sets up the possibilities, denies potential alternates, in essence, it controls the practical and experimental situation. Here is both the glory and the defect of real science.

Obviously, however, the story does not end here. Lister now, in a plain and simple fashion, indicates what happened. Listen to both *what* he says and *how* he says it:

> But when it had been shown by the researches of Pasteur that the septic properties of the atmosphere depended not on the oxygen, or any gaseous constituent, but on minute organisms suspended in it, which owed their energy to their vitality, it occurred to me that decomposition in the injured part might be avoided without excluding the air, by applying as a dressing some material capable of destroying the life of the floating particles.[19]

This is an absolutely textbook case of conceptual richness. What had seemed hopeless in terms of one paradigm seems eminently plausible in terms of another. Note how he refers to Pasteur's researches. They "had shown"—i.e., Pasteur had *demonstrated*, it was *fact*—that the minute particles were the causative agents. Lister evidences no hesitation at all about Pasteur's hypothesis; indeed, Lister apparently does not even consider Pasteur's notions to be a *hypothesis*. Pasteur had hoped, in the 1861 paper, that his ideas were no longer controversial, but had earned the status of "scientific knowledge"; Lister in 1867 certainly behaves as though they had been accepted and established to that degree of certainty. Moreover, given his acceptance of Pasteur's model of fermentation Lister had a clear idea what to do about suppuration. Note the smooth transition Lister makes between the fact that Pasteur's work was about *fermentation*, while Lister was interested in *disease*; his use in this passage of the term "septic properties of the atmosphere" covers this facile transition between radically different subject matters. Lister clearly has bought the entire Pasteur model presented in the 1860–1861 papers. He points out that the "energy"[20] of these minute particles, their causative powers, results from their *vitality*, that is, from the very fact that they are alive. Thus, the potential treatment to prevent suppuration is one which *kills* the minute agent; if dead, then the globule cannot suppurate, it cannot infect. Upon this principle, Lister sets up a practice using carbolic acid (which you can smell in a bottle of Listerine) "which appears to exercise a peculiarly destructive influence upon low forms of life, and hence is the most powerful antiseptic with which we are at present acquainted."

The antiseptic worked wonders. Incidence of gangrene in Lister's wards went to zero in no time at all. Patients quit dying of postwound and

[19]Ibid.

[20]In this time period, "energy" had quite a different meaning than it does for us. It meant something much more like "living forces," or "biological and mental activity and powers," than the sense of "purely mechanical force" which we presently mean by the term.

postsurgical infections in droves, and Lister immediately got the data out in the medical journals.

But Lister's success, although important, is not the essential feature of my account here. What is important is the fact that Pasteur's theory suggests, it promises, it makes new sets of facts potentially coherent—all of this regardless of whether any of these suggestions, promises, potentials ever pan out. In this way the theory gets a hold on the minds of its listeners, and in a strong sense, coerces them to try it, if for no other reason than that it, as opposed to its competing paradigm, is not *hopeless.* Lister understood clearly about the "weed-seeds" in the air everywhere; what he needed to find was the proper "weed killer" to apply to the fertile plots offered by wounded tissue.

Tyndall's Analysis of Pasteur's Potential

In order to drive this point completely home, I would like to briefly report one final instance of the fruitfulness given in the growing coherence provided by Pasteur's hypothesis. This example comes from a paper read in 1876 by John Tyndall, the noted British scientist. The paper title, "Fermentation, and Its Bearings on Surgery and Medicine,"[21] alone shows its conceptual coherence with Pasteur's work. In this paper, Tyndall gives an account of the medicine of his day, entirely in terms of the analogy between fermentation and disease. What is remarkable is the use he makes of the farming analogy. Several crucial points of the paper rely entirely on his audience being able to model fermentation/disease processes upon farming techniques. Let me cite just one or two instances.

One significant terminological modification is used by Tyndall throughout. He calls brewer's yeast by the name "yeast-plant." The implications are obvious. He can now use terms such as "grows," "lives," "multiplies," "respires" with no further explanation at all. Let me give you a characteristic passage, which explains solely by reliance upon the "living plant" model. Tyndall is here talking about how the yeast plant respires when there is no dissolved oxygen in the beer mash:

> In no other way can the yeast-plant obtain the gas necessary for its respiration than by wrenching it from surrounding substances in which the oxygen exists, not free, but in a state of combination. It decomposes the sugar of the solution in which it grows, produces heat, breathes forth carbonic acid gas, and one of the liquid products of the decomposition is our familiar alcohol. The act of fermentation, then, is a result of the effort of the little plant to maintain its respiration by means of *combined* oxygen, when its supply of free oxygen is cut off. As defined by Pasteur, fermentation is *life without air.*[22]

[21]John Tyndall, *Fermentation, and Its Bearings on Surgery and Medicine,* quoted in Conant, Case 6, op. cit. p. 36.

[22]Conant, Case 6, op. cit., p. 41.

Clearly, this explanation makes no sense at all except in terms of what I have called the farming analogy. Yeasts are plants, and they are sown in beer mashes, wherein they grow, respire, etc., producing alcohol as a by-product of their living processes.

In another place, Tyndall relies upon the farming analogy explicitly to set up a model for the specificity rule: one seed type, one crop type. He sets up the case by supposing that someone is given an unknown powder to plant in a prepared garden plot; once it is planted, up comes a crop of docks and thistles. The experiment is repeated numberless times, and each time up come docks and thistles. What must be concluded? It is inescapable that the powder contains at least the seeds of docks and thistles. But now, says Tyndall, suppose

> . . . a succession of such powders to be placed in your hands with grains becoming gradually smaller, until they dwindle to the size of impalpable dust particles; assuming that you treat them all in the same way, and that from every one of them in a few days you obtain a definite crop; it may be clover, it may be mustard . . . the smallness of the particles, or of the plants that spring from them, does not affect the validity of the conclusion. Without a shadow of misgiving you would conclude that the powder must have contained the seeds or germs of the life observed. There is not in the range of physical science, an experiment more conclusive nor an inference safer than this one.[23]

Tyndall then goes on along these lines, explaining how the size of the particles makes no difference, even if they are of microscopic size. From this point, he moves into the airborne "weed-seed" part of the analogy in explanation of the apparently spontaneous generation of fermentations, spoilages, and disease. The upshot of all this is plain. The farming analogy has struck again, this time as a means to make coherent a whole range of observations, data points far beyond those initially rounded up together by Pasteur's hypothesis that the lactic acid reaction involved a living being.

The Anthrax Vaccine

A final incident will show most precisely the potential coherence issue. Tyndall turns to discussion of splenic fever—anthrax—which was a ravager of domestic livestock all over Europe. The story reads like Lister all over again. Let me give Tyndall's own words:

> In 1861, Pasteur published a memoir on the fermentation of butyric acid, wherein he described the organism which provokes it; and after reading this

[23]Ibid., p. 44.

> memoir it occurred to Davainne that splenic fever might be a case of fermentation set up within the animal body, by the organisms which had been observed by him and Rayer.[24]

Davainne and Rayer had earlier carried out microscopic examinations of the blood of anthrax victims. There they had noticed a great proliferation of transparent, rodlike bodies. They supposed these bodies to be organic, but nothing other than slightly unusual. Certainly the bodies were not necessarily related to the disease. But Pasteur's hypothesis about the role of microscopic globules and their life processes in fermentation suddenly made a potentially coherent concept set which could tie together the anthrax disease and the observations Davainne and Rayer had made about the rodlike particles in the blood. Note that their observations made no sense in terms of the chemical decomposition model. The bodies had no potential role to play in the essential processes of the disease according to the chemical paradigm. On Pasteur's hypothesis, in conjunction with the fermentation/disease model, however, these bodies were no accidental components of the disease. They held the promise of being the *cause* of the disease, if Pasteur's hypothesis were accepted.

To make a long, exciting story shorter, I must tell you that Davainne was correct. The rods *were* the causal agents, and by the 1880s anthrax was under control. Tyndall, at the time of the 1876 paper, did not know for sure what would happen. But he clearly responded to the *promise*, the practical potential, constituted by the coherence Pasteur's hypothesis provided for Davainne's observations. The new conceptual pattern had potential, the old pattern had none; this alone was enough to increase the acceptability of the new ideas.

Other uses of the coherence provided by Pasteur's concepts could be cited, and cases of prospective practical effects multiplied endlessly. But I do not need to keep repeating the point. It is clear that Pasteur's hypothesis made so many different areas of information coherent, it provided such great hopes of practical achievement even as it was first enunciated, that these hopes, when coupled with the existential demonstration constituted by the seeding technique, were enough to sweep Pasteur's revolutionary new paradigm into acceptance in record time. The whole period we are discussing here is a matter of a mere nineteen years—from the lactic acid paper Pasteur read in 1857, to Tyndall's lecture of 1876—a mere eyeblink in the passage of time. Given the normal human mental resistance to new ideas, we can infer from this extremely short period of rebellion against the establishment that Pasteur's notions had overwhelming power, the nature of which I hope that I have succeeded in conveying to you.

[24]Ibid., p. 52.

Chapter 11

Acceptance of Pauli's Hypothesis

INTRODUCTION

The processes involved in accepting the neutrino hypothesis are simple, clear, and straightforward. In the first place, the conservative nature of the hypothesis gave it an immediate prima facie acceptability. As I noted earlier, simply because the neutrino concept allowed the energy conservation budgets to become balanced once again, it had a certain degree of acceptability. But there are two other features of the neutrino acceptance process which need discussion. As before, they involve what I call the conceptualistic issue and the metaphysical issue. My discussion of the former will be brief, and for once, directly to the point. Sorting out the latter, however, will take a bit more time and space.

THE CONCEPTUALISTIC ISSUE

Pauli's hypothesis had the initial saving grace of rescuing the principle of conservation of energy from its apparent sentencing to the junk heap.

This obviously has overtones of what I have heretofore called "coherence," in that the hypothesis makes the previously discrepant data now coherent in terms of the conservation principle. But this is not the sole coherence provided by the hypothesis. In a very rapid time, Pauli's hypothesis was developed into a quantitative theory by Enrico Fermi. This theory made a couple of relatively explicit predictions, but ones which could not be specifically tied down to the neutrino hypothesis alone. But growing coherence was soon to be found. As physicist C. S. Wu notes, "Pauli's postulation of the existence of a neutrino to save the conservation laws in beta decay found equally important roles in π-μ-e decays, μ-capture and some modes of κ-meson decay."[1]

Allen presents a similar analysis of the coherence provided by the neutrino concept: "The concept of the neutrino has been so useful in the theory developed by Fermi to explain the beta-decay process that most physicists now accept it as one of the "particles" of modern physics."[2] But probably the best summing up of the conceptualistic issue is that given by Philip Morrison in his marvelous article "The Neutrino." Morrison states the coherence rationale for the neutrino concept in a plain, straightforward fashion which cannot be improved upon: "We could not give up the real triumphs of the neutrino postulate without tearing the present closely webbed fabric of nuclear physics."[3] Although these words of Morrison's were written more than twenty years ago, they are even more true today. More important, however, they give a clear picture of the coherence-providing role played by the neutrino concept at a time when conceptual coherence was the *only* evidence for its acceptance in the concept system of physics. Morrison speaks very nicely of a "fabric," of a set of concepts which are "closely webbed." That this metaphor is descriptive of a system of concepts such as I have defined is quite clear. Thus, we can easily say that the neutrino proved acceptable simply because it provided a large amount of conceptual coherence. It made sense out of a lot of observations and, most significantly, made those observations coherent with the principle of energy conservation.

I am sure you already suspect that something quite odd is going on here. If you go back over the quotes I have given above, it is impossible to miss the somewhat apologetic, indeed, the slightly defensive tone they seem to have. Allen talks about "most physicists" now accepting the neutrino concept because it has proved so useful. On the other hand, Morrison talks about the possibility of giving up "the real triumphs of the neutrino postulate." Why does he even mention such a thing as having to give up

[1]C. S. Wu, "The Neutrino," in M. Fierz & V. F. Weiskopf (eds.), *Theoretical Physics in the Twentieth Century* (New York: Interscience, Inc. 1960), p. 289.

[2] J. S. Allen, *The Neutrino* (Princeton: Princeton University Press, 1958), p. 3.

[3]Phillip Morrison, "The Neutrino," *Scientific American*, January 1956, p. 60.

the hypothesis? What I am getting at is the problem I raised during my first discussion of the neutrino discovery. In order to do what it had to do, the neutrino apparently had to be an undetectable particle. Thus, the neutrino hypothesis, strictly speaking, looked completely unscientific. Reines and Sellschop describe this problem succinctly:

> Because of the apparent undetectability of the neutrino away from its point of origin, it was widely believed at first to be impossible to verify the independent existence of the neutrino in a logically satisfactory manner. Hence, there were legitimate grounds for skepticism among physicists as the actuality of the neutrino. . . .[4]

Morrison says the same thing in an absolutely delightful fashion:

> But now we are on thin ice. Faced with a failure of energy conservation, physicists refuse to admit it but instead postulate an unseen and perhaps unseeable particle—a little neutral one so cunningly designed that it has no properties other than those which will preserve the laws of conservation. How does this differ from plain failure of the conservation laws?[5]

He goes on to suggest the only possible escape out of the dilemma:

> Until you can somehow trace the missing energy, momentum and the rest, you are merely balancing the books with a fictitious entity. There is only one sure answer to the criticism. The missing energy, the "little neutral one" must be caught.[6]

Allen expresses similar sentiments, although in more restrained language; he then goes on to describe the excruciating severity of the problem:

> Complete identification of a new particle is always aided by an experimental observation of some direct interaction of this particle with known particles. In the case of the neutrino, the search for a direct interaction has extended over a period of about twenty-nine years. . . .[7]

Twenty-nine years? What is going on here? How can scientists live with a concept which refers to an apparently unobservable object?

The answer seems to me to be quite straightforward: The choice is between the neutrino and an incoherent physics. The scientists were

[4]F. Reines and J. P. F. Sellschop, "Neutrinos from the Atmosphere and Beyond," *Scientific American*, February 1966, p. 40.

[5]Morrison, loc. cit.

[6]Ibid.

[7]Allen, op. cit., p. 8.

willing to live for twenty-nine years with a "unicorn," a veritable fairy tale, simply because if they did not, the fabric of physics would be rent, the conceptual system would be incoherent. And that is all there is to it.

What we see here is the fundamental priority that logical coherence acquires in science. Science, in great part, is a conceptual object. It exists in our heads in the form of statements, pictures, rules of inference, concepts and their links, etc. If it begins to look as if this conceptual entity is threatened in its logical integrity, the only possible move is to protect that integrity. Similarly, in reciprocal fashion, if the logical integrity is first-rate and growing, if the concepts cohere ever more nicely together, then that in itself is sufficient to make the ideas in question acceptable. And this is precisely what happened in the case of the neutrino. The ever-growing coherence provided by the concept was enough to preserve its life, even in the face of its violation of the almost-sacred canon of detectability/observability.

But the fact that the neutrino's coherence-providing power was sufficient to spare its life did not mean that the scientists concerned were satisfied. The goal of actually capturing a neutrino, of bringing it back alive, drove physicists into one of the strangest hunts which has ever occurred in science. To put it bluntly, the conceptualist satisfactoriness of the neutrino hypothesis fueled a metaphysical search almost beyond compare. Let me now recount for you some of the details of the chase.

THE METAPHYSICAL ISSUE

The best way to describe the motivating reason behind the thirty-year search for the neutrino is by the simple conditional rule, "If a concept makes sense, then its object must exist." Once we have satisfied the antecedent of this conditional with "The neutrino concept makes sense," then the search for the neutrino itself makes sense. However, as you read about it, the search for the particle will seem to you incredible. The amount of money spent, the man-years of effort, the fantastic amount of subtle and sophisticated technology involved, all these factors point to the overwhelming strength of the physicists' desire to capture a neutrino. The almost unbelievable urgency of this desire seems to me to be the only explanation for the intensity of the chase. As far as the underlying rationale for the desire is concerned, I think the only candidate we can offer is the coherence of the neutrino concept as described in the last section. Let me now relate to you some of the exquisite details of the hunt for the neutrino.

The first thing to notice about the neutrino is that it is almost not there as far as ordinary matter is concerned. In the first place, it has no electric charge. This means that it will not leave a track in a cloud chamber; nor can it be affected by magnetic fields, as in an accelerator. Since the cloud chamber and the accelerator are our most effective tools for observing the world of microparticles, it is pretty clear that we start out on the neutrino hunt with two strikes already against us. But its lack of charge is not the neutrino's worst problem. What finally almost makes the hunt impossible is that it has a very tiny interaction cross section, which is the same thing as saying that it does not very often bump into other things. Morrison puts the case plainly:

> But a neutrino, to interact with a nucleus, must travel on the average through about 50 light-years of solid lead! A shielding wall capable of thinning out a beam of neutrinos would have to be as thick as a hundred million stars. To all intents and purposes neutrinos do not see solid matter at all. Here is the nub of the difficulty. The neutrino is almost uncapturable.[8]

Faced with this situation, physicists initially were, as I noted earlier, quite skeptical about ever demonstrating the existence of the particle. Things at first looked exceedingly gloomy. However, as theorists kept working with the concept, other possibilities began to open up. What happened was the continuation of the growth of the neutrino's conceptual coherence. The hypothesized particle began to crop up in rather diverse areas of elementary particle physics. The neutrino, in fact, soon leapt the walls, becoming no longer tied entirely to beta decay processes—those involving what Fermi called the "weak force." It also came to function in nuclear and subnuclear notions. It was at that point, during the early and midfifties, that some new reactions, reactions with a potential for indirectly observable neutrino interaction, were dreamed up in theory. Although they are a bit complicated, you will need to roughly understand these theories in order to appreciate the incredible details of what happened next.

Cosmic rays are heavy, fast, extremely high-energy particles which come streaming into our vicinity from deep space. Usually they are protons, the positively charged nuclear particle. When the rays encounter earth's atmosphere, they often collide, in an extremely violent way, with nuclei of the various atmospheric gases. In the collision the atoms come apart in all sorts of chunks. One of the subnuclear chunks is called a pion (π). The pion is somewhat unstable, and it soon decays into a muon (μ). The muon in turn decays into a very energetic neutrino. Finally, of ultimate importance, there are lots and lots of these neutrinos; millions go through every earthly square inch every day. Thus, even though neutrino interaction with matter is rare, it can be expected that one out of every

[8]Morrison, op. cit., p. 61.

10^{20} will interact with a suitable substance. According to this, the large number of cosmic/atmospheric neutrinos is great enough to make up for that rarity of interaction. The problem remaining, however, is to set up a detector which can interact with the extremely high-energy atmospheric neutrinos. Here is the interesting part.

Neutrinos of the high-energy sort can be expected to interact with the nuclei of the soil, terra firma, rocks, etc., producing muons. Muons are somewhat rare birds, but there are enough of them around from other sources that observing a muon does not count as observing a neutrino. But a clever ploy solves this problem. Muons are relatively easy to shield. That is, they interact fairly easily with normal matter, and thus can be stopped by a mile or so of standard dirt and rock. Thus, if we could put a detector in a place which had a mile or so of dirt all around it, then we should be shielded from the typical standard muon. But this mile or so of dirt is as nothing to the neutrino. It passes through the shield as if the shield were not there. But if there are *enough* neutrinos, some stray ones will interact with the dirt. Presto! A muon appears, in a place where it should not be. And the occurrence of a muon where it should not be is taken as evidence for the existence of the neutrino, since only the "little neutral one" has the power to produce a muon where it should not be.

Now, where in the world can one find a place which is surrounded by a mile of dirt on all sides? In a South African gold mine, of course! Let me allow the researchers themselves to describe the situation:

> All that remained was to find a suitable site and then to build, install and operate the equipment. A worldwide search resulted in the selection of an excellent location in the East Rand Proprietary Mines near Johannesburg. There space for a laboratory was specially hewn out of solid rock 10,492 feet below the surface. The laboratory consists of a large entry area, which contains the bulky electronic equipment and a tunnel 500 feet long and eight feet in diameter, where the detector elements are installed.[9]

Hunting for neutrinos, as you can see, is no easy business. But the description is not over yet; we have not taken a look at the detector equipment. Here is what it is like:

> The detector equipment includes five miles of cable, 4,400 gallons (16 tons) of scintillating liquid, 144 photomultiplier tubes and a structural framework 120 feet long. The logistics of manning the operation are illustrated by the fact that members of our group have logged a total of 500,000 miles of travel between the U.S. and South Africa in the past two years.[10]

[9]Reines and Sellschop, op. cit., p. 46.
[10]Ibid.

What happens is this. The muons slam into a huge tank of mineral oil, which has been treated with a special chemical. When the muons interact with the oil they emit tiny flashes of light (scintillations) which are undetectable by human eye. But the tanks are surrounded by banks of photomultiplier tubes, which are able to detect the flashes. When the muon triggers the photomultipliers in just the right sort of temporal and geometrical sequence, the detector counts the flash as a neutrino "event."

The huge, complicated device was first turned on in October 1964. It soon registered its first *sister* muon, the type which came directly from a cosmic ray collision in the atmosphere (those which are formed directly in the dirt and rock are called *daughter* muons, and these are the significant ones). Let Reines and Sellschop tell the conclusion in their own words:

> Then on February 23, 1965, the detectors recorded a muon that had travelled in a horizontal direction—the first "natural" high-energy neutrino had been observed! Since last February we have observed in our equipment some 80 sister muons and 10 daughter muons.[11]

Ten events! All the work, the trials and tribulations, expense, time, equipment for ten events. Behavior such as this is strange, to say the least. Certainly the ordinary person in the street must raise an eyebrow at all this, and ask why? Why go to such lengths to try to observe such a shy creature as the neutrino?

The answer could not be simpler. These experiments have shown that the neutrino has to exist, given how well the concept fits in, coheres with, the physical paradigm. Morrison, again very nicely, sums up the case. Writing nine years before the South African mine experiment, he describes the stakes:

> But suppose the experiments do not work? Suppose no neutrino counts are seen? *The logical chain is pretty tight*; the defeat would mean to many that energy conservation had at last really failed us, or, almost as bad for our theories, that a reaction was not accompanied by its inverse among the fundamental particles [i.e., that there was not a symmetry in fundamental particle interaction]. We should be loath to accept either of these conclusions. . . . It will be far better if the patient experimenters are successful, and if their scintillator clearly displays those few oscilloscope traces each hour which will mean that the fugitive neutrino has been caught at last.[12]

Those ten events are ultimately of vast significance. They justify meta-

[11]Ibid.
[12]Morrison, op. cit., p. 68.

physically a belief which had been held for over thirty years on epistemological grounds alone. Coherence is sufficient, it will provide evidence enough for the use and belief in a concept. But the metaphysical goals must also be sought for; the objective reality of the referent of the concept must be evidenced in some way, no matter how indirect. Thus, while the coherence of the neutrino concept was sufficient to grant its acceptance in modern physical theory, it was not enough to dissipate the uneasiness felt by most physicists in the face of the credulity-straining undetectability of the particle. Those ten events clinched the case. The existence, the metaphysical objectivity, of the neutrino meshed with its conceptual power, and physics could then march on, deeper and deeper into its exploration of the structure of the natural world.

Chapter 12

Conclusion

REVIEW

I began this work by talking about the three main divisions of philosophical analysis: metaphysics, epistemology, and logic. I went on to claim that many problems in science are not really "scientific" problems, but rather are "philosophical," according to the three categories. Some scientific puzzles concern questions about what kinds of objects the universe is populated with. Others concern issues and debates about the role our knowing processes themselves, perception and observation for example, play in gaining scientific knowledge. Finally, inextricably wound up in the first two, logic always asks questions about the formal properties of our reasoning: is the concept "consistent"? Does the prediction follow "deductively"? Do the two theories "cohere"? So went Chapter 1.

Unfortunately, as you may recall, Chapter 2 followed Chapter 1 as the bleary night follows the clear day. Logic, its formal processes, and the notions of a deductive system all came to confront you. Probably of most importance, you collided with the '→,' which was to plague you to the very end. Conditional reasoning, the "if, thens" of science and ordinary life, proved a bugbear of large proportion. But perhaps you learned something from your encounter with it.

In Chapter 3 I introduced you to the heart of this book, notions which were to be threaded into all subsequent discussions. The first of these notions was "the goals of science." "Explanation" and "prediction," I argued, have always been the two most prominent answers to the question "What are the objectives of science?" Slugging it out over the centuries, while the pendulum was swinging back and forth, thinkers about science have tended to come down hard on the side of one goal, somewhat to the exclusion of the other. I had at first wanted to say that both goals were equally significant, but as I wrote about how the ideas of Lavoisier, Pasteur, and Pauli came to be accepted, I found that I too in fact tended to push one goal rather more than the other—a subject which I will discuss more fully below. But even though I find myself safely ensconced in the camp of those who emphasize one of these goals to the neglect of the other, I still find myself wanting to say that the two goals are equal. I think they probably are equal, even though I have not succeeded in presenting them as such.

The second notion introduced in Chapter 3 was that of "a conceptual system," with all its variations—"paradigm," "belief structure," "theory," "hypothesis," and so on down the line. You would surely agree that there has been quite a bit of talk about conceptual systems since that first introduction. Probably the conceptual system is the most fundamental idea of the whole book. I claimed that the conceptual system is of prime importance in the education of science students; that it is, ultimately, the guardian and leader of scientists in their gropings toward new hypotheses/discoveries; and finally, that it is the tradition against which scientists rebel when a discovery is in hand. That is, the growing periods in science involve a kind of conceptual warfare between ideas.

Still, I also tried to show, from time to time, how practical affairs, the need for prediction, control—technology in short—always nagged at the prevailing conceptual system, biting its heels when it moved too slowly; caroming off it when it moved too fast. Reason demands that practice always conjoin to theory; usually it never completely does. From this tension, just as from the tension between vying conceptual systems, progress—or, at least, movement of some sort—ensues.

We spent a lot of time in Part Two looking at three case histories of scientific discovery, trying to figure out how Plato's dilemma could be resolved, and discovery accomplished. Along the way we took side trips into the problems of reductionism as well as a quick look at the role of the conservation laws, and of course, we constantly and consistently strayed off the path into points about the priority role of the mind—its concepts, perceptions, notions, ideas, what-have-you—in the scientific processes of observation, hypothesis, and discovery.

In Part Three we have taken a look at how theories come to be established doctrine, at how the scientific community finally decides to accept some new idea, which always involves the simultaneous decision to discard some redundant chunk of mental entity, some superseded conceptual system. I pointed out the various schools of thought which have taken positions in the issue of how to accept theories, and then promptly came down on the side of two of them, the metaphysicians and the conceptualists. Here again, I think that I am guilty of a sin of excess—too much enthusiasm about and emphasis on my own beliefs, rather than an objective presentation of the material. So be it. I am sure that my fellow members of the theory of science community, all my logician/philosopher/historian-of-science colleagues, will submit my views to the requisite trial by fire. Not all my ideas will survive such a trial; but I am quite confident that some of them will survive and grow.

MY CONCLUSION: FIRST POTSHOTS AT MY IDEAS

I want now to analyze the exact nature of my sin of excess, and to carry out the analysis in terms of the goals of science. Explanation and control are, to my mind, the two goals of science. The one relates most intimately to our minds, whereas the other is ultimately connected to our bodies and lives. At one time or another, the history of the various sciences exhibits a science which can do one of these, but not the other. Often, however, a science can do both. Some theorists, though, have argued most particularly for the significance of the one, and the denigration of the other. Plato, for example, thought that explanation alone was satisfactory enough. The modern psychologist B. F. Skinner believes that theory has no claim upon our scientific activities; control and prediction suffice. I argued that *both* are necessary elements of any healthy, whole science. But then I spent an inordinate amount of time talking about, or seeming to talk about, the sufficiency and significance of the drive to explain, of our motive to involve our minds in understanding Nature in such a way as to make sense of Her. I have concentrated far too much upon the conceptual system, the central element of the rationalist position I hold. But there is a strong reason for this emphasis which I shall now confess; and moreover, there is a saving grace which can fall upon my analysis of the acceptance phase, a saving grace sufficient to make something of objective value out of my excess.

In the present century, the pendulum has swung too far toward the empirical pole of science. Practice, technology, and prediction and control lie too near and dear to positivism, the spirit of our recent age, for this extreme swing of the pendulum to have been prevented. But the swing appears to me to have reached its maximum; the weight of thought is now poised, ready for the rebound into a much more conceptualistic/rationalistic perspective on science. I think that this present work (of which you can now see the light at the end of the tunnel) is just another indication of the coming direction of the swing. After all, my professional teachers, each and every one, were either students of positivists or living, breathing positivists themselves. And students most always end up in some sort of rebellion against their teachers and their teachers' ideas. Plato and Aristotle are paradigms of this natural law. I think that some real progress toward understanding science was achieved by the positivists. But it is now time to move on again, with their fresh synthesis as the jumping-off place. Of course, as any of them would tell you, this book represents quite a long jump. But such a jump could only be expected.

Now for the saving grace. As I worked through my analysis of theory acceptance processes, I began to notice that I had settled down into an analytical rut: Coherence and existential verification were the two features, respectively epistemological and metaphysical, which characterized the main currents of the process by which a scientist's hypothesis came to be accepted. Thus, I argued, when the coherence of the new conceptual system satisfies our minds, we must engage ourselves in capturing one of the beasts which the concept identifies. Again, the mind's eye leads the body's eye (and hand, too, of course) to the snaring of the creature named by the concept, even when the chase is as arduous as that for the neutrino.

But I got a hunch while trying to work all that out: Somehow, some way, the coherence criterion is related to the explanatory goal of science; and similarly the metaphysical quest is related to the practical objectives of science. I have not in any way sorted this hunch out, but I think I might be able at least to say one or two sensible things about it.

Philosophers and logicians talk a lot about the "explanatory power" of hypotheses. But I think that explanatory power, and what I here called conceptual coherence, are so inextricably related that it does not make sense to talk of them independently. If this be the case, and I suspect it is, then the growing coherence of a new idea is at one and the same time the growing satisfaction of the goal of explanation. This would explain the inordinate significance attached, for example, to the growing power of the neutrino concept to make coherent diverse areas of physical theory. The neutrino, on this analysis, is acceptable simply because it quite satisfies the scientific impetus toward explaining nature. But we have predisposed ideas about the relation between coherent concepts and the real, physical mechanisms of the world. Mainly, if the concepts cohere, then the object, the purportedly physical mechanism which the concepts name, *must* exist, it must *really* be out there. The final aspect of my hunch lies now before us: The metaphysical quest, the drive toward verifying the existence of the mechanism alleged by the concepts, is fueled by the desire to satisfy the goal of control.

This is a rather tenuous insight, if it is an insight at all. But I can at least say this in its justification. When we have verified the existence of the hypothesized object, then we have caught one, we have brought it back alive. To be brutally blunt: We have *controlled* it, we have manipulated it, we have brought it under our power. Thus, the ontological quest is simply the manifestation of our desire to accomplish the objective of control.

Another explanatory point also comes from this. When scientists are skeptical about accepting an apparently undetectable, unverifiable mechanism, they are most likely afraid that this object could never be *controlled.* If you cannot even verify that it is out there, then how could you ever hope to *use* it, manipulate it, in any significant sense at all? If this is the situation, then the unverifiable object, apart from its conceptual fruitfulness, is absolutely useless. Naturalism requires that the object be natural, i.e., physical like us. If we cannot verify its existence, then maybe it is not physical. And if it is not physical, then it can never be scientific, since it cannot be used in satisfaction of the control goal of science.

So runs my insight, my hunch, about the relation between the goals of science, and the metaphysical versus epistemological aspects of accepting theories. I hope I am right, but maybe not.

It is time to end. My final hope is that you have come to a bit more appreciation of science as an activity which is intrinsically human. Human beings *do* science, and they often act like normal people while doing it—how could they do otherwise? But sometimes, when science is at its most rational, scientists perform better than people often do; and for this reason it is valuable to study science. I hope that you can now agree with this reason for studying science. If not, then I have failed utterly. I hope not.

Glossary

In this glossary I try to define very informally some of the unfamiliar terms that you run into in the text. Although the precise wording of the definitions here may not exactly correspond to the usages in the text, the meanings will be close enough to be of real value to your understanding.

Aesthetics A branch of philosophy which studies ideas and standards of beauty. Thus, "Are some scientific theories more beautiful than other scientific theories?" is an example of an aesthetic question.

Algorithm A recipe or formula which will produce some desired result, often with no knowledge of why and how it works required on the part of the user. Thus, "On cold mornings always turn the car lights on for thirty seconds before trying to start the car" is a very simple algorithm for success in cold engine starting. Algorithms are usually formulated in statements called conditionals, that is, statements in the form "Whenever thus and so occurs, then do such and such," which list the conditions under which certain things should be done.

Conceptual system A group of rather precisely defined and clearly delimited ideas, which are tied together in an interlocking way. The logic of the interlocking relations is somewhat like the logic of a wardrobe; that is, if you change the style of one element, then all the others must also be changed in order to keep the set all in the same style. See Hypothesis, Paradigm, Theory, and Tradition.

Epistemology A main branch of philosophy, which studies things having to do with human knowledge, the processes involved in knowing, and the limits and boundaries of these processes. See Philosophy.

Hypothesis A proposal that a new concept (or a small conceptual system) is correct, and thus deserves its own proper place in the current paradigm. See Paradigm and Conceptual System.

Logic A main branch of philosophy, which deals with the patterns and procedures involved in thinking and reasoning. See Philosophy.

Metaphysics A main branch of philosophy, which studies things having to do with existence and reality. Thus the question "Do angels (or atoms, or souls, or psychic forces) really exist?" is a metaphysical question. See Philosophy.

Paradigm A wide-ranging conceptual system which includes many small conceptual systems (theories) as components. The whole interrelated structure is focused and organized around a set of scientific accomplishments—a successful series of experiments, for example. A paradigm is what is taught in the textbooks, in the schools, and in the graduate schools. See Conceptual System.

Philosophy An attempt to provide answers to some ultimate, very basic questions. Philosophy does not begin until humans reflect upon their experiences and activities, and start to ask questions like "What kinds of things can I know, and how do I know that I know them?" and "What kinds of things really exist in the universe?" and "Is there any best way I can set up my thinking processes in order to make them as accurate as possible?" Each of these questions leads to a main category of philosophy—respectively, epistemology, metaphysics, and logic.

Theory A conceptual system which is limited to a more restricted subject area than is a paradigm and, usually, is not quite so strongly believed in as would be a paradigm or tradition.

Tradition Although wider in range, a tradition includes a conceptual system which functions much as a paradigm does. But a main difference between it and a paradigm is that a tradition has endured for some good-sized piece of time, and thus has an historical, institutionalized life of its own. Members of a tradition tend to define significant aspects of their lives in terms of the tradition. Traditions tend to eat up and digest related conceptual systems, as well as producing their own in their own terms. The American political tradition is an example of a political tradition. Roman Catholicism is an example of a religious tradition.

Index

Index